SD-WAN 1:1

The What, Why and How

List of authors

Tejbir Batth
Matt Berry
Joe Harris
Yves Hertoghs
Oliver Kupke

Cliff Lane
Rosa Lear
Joe Lee
Stephen Lynn
Rohan Naggi

Gerd Pflueger
Sajjit Rajagopal
Martin Rausche
Dave Twinam
Ray Wong

1. The Software-Defined Vision for the WAN — 5
Introduction — 6
Allofu, Inc.—Everyone's Story — 7
Why Does the Wide Area Network Need to Evolve? — 9
What is SD-WAN? — 10
SD-WAN Use Cases — 11
The SD-WAN Vision: Network Edge — 21

2. Components — 27
Solution Components — 28
Placement of Components — 34
Communication between Components — 35

3. Getting Started — 39
High Level VMware SD-WAN Deployment — 40
Zero-Touch Provisioning — 45

4. Topologies — 51
Overlay Options — 52
Overlay Topologies — 60
Deployment Models — 69
Site Topology and Redundancy — 74

5. Application Performance, Routing, and Cloud Access — 87
Application Performance—Dynamic Multi-Path Optimization — 88
Routing — 94
Business Policy — 103
Optimized Cloud Access — 113

6. Security — 123
Security—VMware SD-WAN — 124
Security of VMWare SD-WAN Solution Components — 129
Inspection of User Traffic — 131
VMware SD-WAN Segmentation — 142
Securing SD-WAN Deployment — 148

7. Migration — 155
From Traditional WAN to SD-WAN — 156
Routing Consideration During Migration — 159
Migration Workflow — 167
SD-WAN Migration Journey — 174

8. Service Provider — 179
Service Provider Topology — 180
Service Provider Routing — 186

9. Integration — 197
Operations — 198
Troubleshooting — 202
vRealize Network Insight for SD-WAN — 222
API — 223

10. SD-WAN in the Bigger VMware Picture — 231
SD-WAN in The Virtual Cloud Network — 232
VMware SD-WAN and vCloud Network Function Virtualization — 236

11. Appendices — 243
Acronyms — 244
Glossary — 246
Additional Resources — 253

1. The Software-Defined Vision for the WAN

Introduction

by Sanjay Uppal, former CEO and co-founder of VeloCloud, and current VP and GM of the VMware SD-WAN by VeloCloud Business Unit.

I am often asked two questions by interviewers, analysts, and customers. The answer to the first one: "Why do you like The Big Lebowski so much?" is always "all of life's answers are contained in this movie." The second question is, "Why did you choose to take on SD-WAN?"

Well, why not? When my fellow founders (Steve Woo and Ajit Mayya) and I were brought together, we had a vision of solving a major networking problem that would have an immediate and far-reaching impact on how people and businesses worked. Having been in the networking space for most of our professional lives, the three of us decided to tackle improving connectivity with a technology that would make it faster, easier, and better to configure and manage but also support this thing we kept hearing about: The Cloud.

Every new company and product were being named for the cloud. How could we take advantage of this trend and who would buy the solution? Originally, we thought it would be consumers, because who doesn't want their at-home applications to run faster with a stable, reliable, connection? But then we thought bigger and decided that the enterprise would be a better fit. And so, we were set: Enterprise + Networking + Cloud + Fast + Simple + Evolving = SD-WAN.

The secrets were in the design of the solution: the fact that it wasn't hardware-based (it is software-defined) and that it was easy to deploy and consume. And that's what we made: a transformational technology that will challenge how you currently think about your traditional network and offer you a solution that has no end in sight to the capabilities it can perform and enable.

This book was born entirely out of the minds of our front-line team members—the engineers who sell, deploy, and engage with SD-WAN on a daily basis. They understand your pain points, your network as it is today, and what it should look like tomorrow. Most importantly, they understand better than anyone how to make it all happen. So, let's get reading, and to quote The Big Lebowski as it relates to SD-WAN: "It really tied the room together."

Allofu, Inc.—Everyone's Story

Throughout this book, we will follow the evolutionary story of a fictional company, Allofu, Inc., an embodiment of the organizations seeking a solution to, or improvement of, their existing network infrastructure into today's age of digital transformation. The story will be told from the vastly different perspectives of two Allofu employees who share the same goal of improving the business by meeting its technical needs. Each section of the book will begin with the problem faced by either Alvina or Rodney, followed by the technical solution and best practices, and close with the tactical actions taken by Rodney.

Alvina is the Chief Technology Officer, responsible for the overall networking strategy and direction of the company, return on investment of all networking-related solutions, and business goal achievement.

Rodney is the network architect, responsible for the organization's networking and infrastructure-related activities and tactical implementation of the solutions outlined by Alvina.

Let it begin...

Allofu, Inc. is an international company with its headquarters located in the Americas, a regional distribution center in Europe, and multiple offices and branches located throughout both territories. It offers 50 different products to its customers and handles thousands of calls every day, as well as receiving thousands of hits to its website on a monthly basis.

In order to support its business at scale, Allofu has data centers in both America and Europe hosting most of its applications, and a number of applications in the cloud. Allofu's current infrastructure relies heavily on MPLS to support its network and business needs. Overall organizational goals include expanding its branch footprint in both America and Europe, opening new sites in Asia, and ensuring that all offices, data centers, and branches are connected to each other. Additionally, Allofu is evaluating the acquisition of another company and anticipates challenges with integrating the networks of the two organizations.

With an expanding footprint, increasing customer needs, and pressure to move applications to the cloud, Allofu is hindered by its existing WAN network as it strives to achieve its growth and customer satisfaction goals. In order to continue to expand and support its employees and customers, Allofu must rethink its existing network.

> **TECH TIP**
>
> The use cases, best practices, and tactical recommendations made throughout this book are applicable to the majority of organizations. Whether the organization has 5 or 50,000 sites, the solutions can be adapted to specific cases and usages. For instance, throughout the book Allofu will follow a growth trajectory, adding new applications, new regions, and new branches to its topology. In the case that a real-world organization might not follow that trajectory, the topology and migration strategies are still applicable but on a smaller scale.

Why Does the Wide Area Network Need to Evolve?

Businesses today rely on their WAN network to connect branch and remote locations, and to access private data centers, SaaS applications, and various resources in the cloud. The availability and costs to operate and maintain the WAN have increased along with the associated complexity of managing this critical service. Leveraging the internet for connectivity is perceived to not be reliable enough, particularly for latency-sensitive traffic such as voice and video. Additionally, security is always a concern when sending and receiving traffic over the internet.

Driven by the growing number and requirements of modern applications, business needs to be more agile than ever and this often includes the ability to make changes to the WAN. Applying changes to a traditional WAN is generally considered challenging and complex; both the careful planning required to coordinate the necessary changes based on the users' requirements, and the need to comply with maintenance windows, can negatively impact business agility.

Businesses are rapidly moving data and applications to the public cloud, often without taking into account the changing WAN requirements associated with such a move. With current WAN architectures, this move means backhauling all the traffic from the branch to the data center before it is forwarded to the internet. This approach negatively affects other application flows on the WAN and is also an inefficient use of bandwidth.

Application performance over the WAN has always been crucial to the needs of the business. Traditional WAN offerings have been unable to fulfill modern application service level agreements (SLAs) and bandwidth requirements. Even utilizing multiple connections brings its own set of challenges. And this problem has been exacerbated by the aforementioned move of applications and data to the cloud.

When one asks, "Why does the WAN need to evolve?," it really comes down to this: Superior business agility, improving application performance, effective support for data and applications on the public cloud, simplification of the WAN.

What is SD-WAN?

Software-Defined Wide Area Network (SD-WAN) is the application of software-based network technologies to the WAN to more effectively transport network traffic.

SD-WAN dynamically utilizes all available connections (private circuits, broadband, DSL, 3G/4G LTE, etc.) to find the optimal delivery path for traffic across the entire network, thereby delivering an optimal user experience regardless of location or access type. As different applications require different characteristics for transport, SD-WAN dynamically steers traffic to the best available connection based on real-time network conditions.

In industry-accepted terms, to be categorized as SD-WAN, a solution must contain four specific characteristics:

- The ability to support multiple connection types, such as MPLS, Ethernet, and higher capacity LTE wireless communications.
- The ability to carry out dynamic path selection for load sharing and resiliency purposes.
- A simple interface that is easy to configure and manage.
- The ability to support VPNs and third-party services such as WAN optimization controllers, firewalls, and web gateways.

Introducing VMware SD-WAN by VeloCloud

VMware SD-WAN by VeloCloud delivers a highly reliable and secure application-centric service for even the most latency-sensitive applications, independent of the underlying links. This is achieved by leveraging a simplified cloud-based platform that delivers the required business agility, performance, and simplicity. VMware SD-WAN ensures the secure delivery of traffic across various transports including the internet. It uniquely delivers simplified management with elimination of traditional CLI-based configuration and monitoring.

SD-WAN Use Cases

SD-WAN lends itself to solving multiple use cases, too many to exhaustively discuss in this book. However, after several years of deployments, six use cases have emerged that are common across the majority of those customers who use VMware SD-WAN. In this chapter, those six use cases are discussed, showing how each impacts both the business side (told from the perspective of Alvina) as well as the technical side (told from the perspective of Rodney) of organizations.

Deploy Multiple Links (MPLS/Broadband/LTE)

Alvina is concerned with the cost required to manage and deploy branch locations. Obtaining the correct margins to remain profitable requires that all assets are optimized, which often leads to reducing costs. Alvina's sales leadership team has frequently complained that the "network is slow" and is leading to customer dissatisfaction and lost revenue. Voice is a business-critical application and several branch locations have experienced sporadic outages in voice traffic during peak hours. The private circuits providing connectivity to the branch locations are costly and comprise one of the larger monthly costs of operating the branches.

Rodney is under constant pressure to determine the source of end-user complaints. It has become increasingly difficult to diagnose issues using the existing Network Management System (NMS) and the various log files available to him. The cost of the private lines from his providers is high and there is no budget for additional circuits to support increasing network demand. In addition, Rodney is under the assumption that the provider infrastructure should protect the critical VoIP application, which has not proved to be so.

It is very common today that branch locations have a single active link for all connectivity as shown on the left side of Figure 1.1. This link is then used for all traffic in/out of the site, regardless of the business criticality or application needs. This link is commonly a private circuit (often MPLS) but may also be utilizing a broadband connection.

Depending on the cost associated with a private circuit, it is sometimes financially impractical to have multiple links. This often means that in the event of network congestion or outage, the business can be seriously impacted. It can be of significant value to an organization to have the ability to cost-effectively add additional links to support both bandwidth-hungry and latency-sensitive applications.

VMware SD-WAN brings application performance and availability as well as visibility and control to single or multiple links as shown on the right side of Figure 1.1. These links can include existing private leased lines, broadband, Ethernet, and increasingly, LTE links where limited connectivity options exist. It offers the ability to rapidly respond to network events, and optimize throughout the network while being conscious of the critical network requirements of each application.

The ability to define network policy which reflects the business importance, as well as the specific requirements of an application, are critical. One of the benefits of VMware SD-WAN is the ability to quickly and easily apply business policy to an application, based on the needs of a particular organization and the characteristics of the application.

Critical business applications for a given business, such as VoIP or a SaaS-based application, can be assigned to the appropriate business policy, either from pre-defined profiles or through a user-defined process by the network engineer. This business policy can include which links to use per application. Furthermore, if sub-optimal network conditions are detected, VMware SD-WAN has the ability to dynamically steer traffic on a per-application and per-packet basis to yield the best performance based upon the needs of the organization.

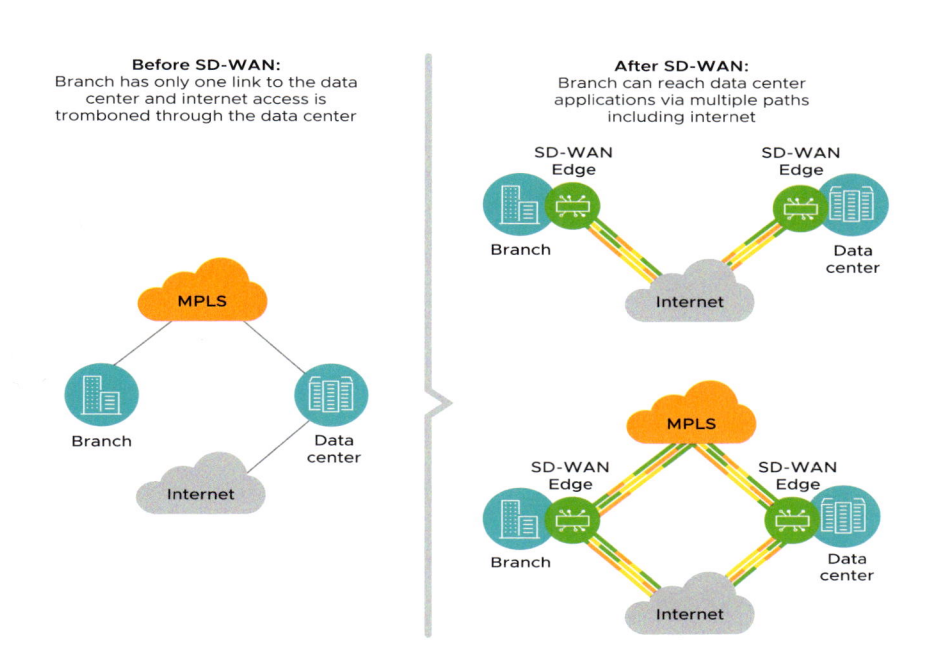

Figure 1.1: VMware SD-WAN Edge deployment with single and multiple links

Alvina is optimistic that the application prioritization capabilities of SD-WAN would ensure that the business-critical applications for her business could provide the necessary application performance that her sales leaders and customers require. For her most cost-sensitive sites, a single link would be sufficient when using application prioritization. For the busiest branches, a cost-effective internet link (or LTE for the most remote locations) would provide the ability to increase bandwidth while also benefitting from the application prioritization enhancements of SD-WAN.

Rodney is concerned because he has been down this road before in making improvements to the network with a "revolutionary" new technology. When that network was deployed, it was positioned as the network of the future and that's not quite working out as one would have hoped. Adding links is adding cost, so how does it get any less expensive? Is the internet reliable enough for real-time applications such as voice? How does SD-WAN help him gain an understanding of his network and what is causing the complaints about network performance? Won't adding more overhead via an overlay add more complexity and reduce his visibility?

Simplify and Accelerate Branch Deployment

Alvina needs the organization to become more agile. It is the digital age, but Allofu IT appears to be rowing upstream and unable to keep pace with the needs of the organization. Alvina knows that her IT team has to become more agile and she feels that perhaps a technology such as SD-WAN would give them the toolbox they need. Cost optimization does not include adding more and more engineers to the networking team, and it is critical to do more with less. The speed of business growth means that if you cannot respond faster than the competition, do not bother trying. The network seems to constantly be an obstacle to both speed and agility. It seems like nothing has changed for years.

Rodney is frustrated by the constant drumbeat of "do more with less." It's not as if nobody is trying, but it cannot be helped if it takes several weeks to provision new circuits. Rodney has worked hard to develop scripts to make it faster to roll out new sites, but deployment time is increased when someone from the IT team has to go on-site to troubleshoot problems. In addition, there are demands and coordination required from the IT security team, which is not exactly Rodney's idea of responsive.

The ability to rapidly and easily deploy new branch locations has long been a challenge for organizations. The time-to-market aspects typically begin with the ordering and provisioning of private circuits. It is not uncommon for this process to require weeks, or even months, especially for international connections, and typically involves a third party, i.e. the telecommunications provider of the private circuits.

Further challenges to branch deployment commonly include the need to schedule site visits from a trained IT team to physically manage the installation and activation of a new location. This process also commonly requires coordination with other teams within IT to meet the needs of the organization, as well as finding maintenance windows to make changes to the network infrastructure. What organizations increasingly need is to significantly shorten the time required, as well as to greatly simplify the process.

VMware SD-WAN can leverage any public or private circuit, including LTE, which can be provisioned much more rapidly. VMware SD-WAN also offers the ability to deliver zero-touch provisioning of new network endpoints. Physical and virtual appliances automatically authenticate, connect, and receive configuration instructions once they are connected to the internet. This activation can be completed by any individual that can physically connect the necessary components. Once activated, the network engineer can remotely perform all configuration, monitoring, and troubleshooting.

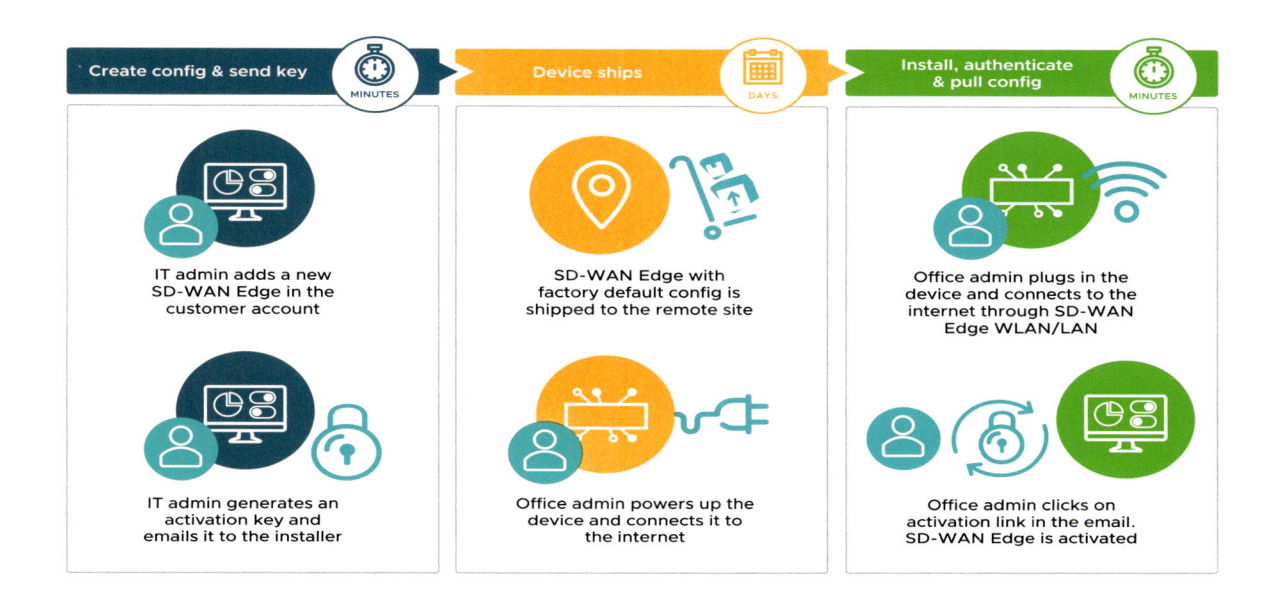

Figure 1.2: VMware SD-WAN Edge zero-touch provisioning

Alvina thinks this process would have an immediate and positive impact on her travel budget as there would be fewer truck rollouts along with Rodney's penchant for steak dinners when he is on the road. If SD-WAN truly allows for internet links and can protect her business applications, then why not just call the local broadband provider or just use LTE to shortcut the weeks of waiting for a circuit to be installed?

Rodney remains unconvinced as the old adage of "if it seems too good to be true, it probably is" still rings true to him. From Rodney's perspective, this application prioritization had better work, otherwise, there is no chance of usable VoIP traffic. The idea of not going on-site for every install is pretty appealing, but he really needs to see it working to believe it.

Access to IaaS and SaaS

Alvina needs to figure out how to reduce the cost to operate her data centers. Data centers cannot compare with the time-to-market promised by the cloud, and the costs to operate a data center do not make sense now that cloud has been proven to be ready to provide support for business needs. SaaS applications have been in the market for some time and have resulted in a considerable positive ROI.

Today, some of Allofu's branches get better performance than others, but according to Rodney it is a function of the number of hops back through the provider's network and that is not something he can control. There has been a recent rash of incoming bills from public cloud providers and from engineers that decided to spend money first and ask permission later. They had good intentions, but having developers run around spending money would undermine everything the cost-optimizing efforts could return.

Rodney's head hurts with all the incoming demands and he is getting tired of having to constantly repeat "it wasn't me!" and "Oh, by the way, they did not consult anyone within IT first!" when finance shows up waving a giant bill the development engineers ran up. It would also be nice if employees would stop coming to him because they chose to live in some remote technologically backward campground that a sane provider would never expect something other than a squirrel to access the internet from. He's also tired of saying "yeah, it's SLOW and it's not my fault."

The need to leverage cloud infrastructure resources or applications has become universal. This may be anything from accessing workloads running on a public cloud (known as Infrastructure-as-a-Service (IaaS)), cloud-based storage, or increasingly Software-as-a-Service (SaaS) applications. While these resources are commonly being utilized, they are rarely being accessed from the branch locations in an efficient manner. In the typical use case of traditional WAN, the traffic is commonly backhauled to the data center for security policy reasons, and this leads to an increase in latency and negatively impacts the user experience.

VMware SD-WAN offers a powerful solution to optimize access to cloud resources as well as global access to SaaS across the globe. Utilizing broadband links, VMware SD-WAN allows branches to connect dynamically to these locations, providing direct access to the closest required cloud resource, and thereby avoiding unnecessary backhauling of traffic to the data center.

Alvina sees clear advantages for Allofu to move more of their applications to the cloud. Alvina also needs to address the shadow IT that has sprung up to fill a need they can see. Both IaaS and SaaS will increasingly have a prominent role in Allofu's future needs, and it's promising that the VMware SD-WAN solution appears to provide potential performance enhancements which would make her team more productive and hopefully complain less. The key for Alvina is control of how the Allofu end-users consume those services.

Rodney is frustrated that some of his colleagues have taken it upon themselves to go spin up instances on public clouds without bothering to request assistance from IT, much less to verify if these rogue decisions are even business compliant! Rodney is interested to find whether the VMware SD-WAN offering can improve performance for end-users accessing IaaS and SaaS. Rodney likes anything that makes his life easier, and that includes fewer users complaining "the network is slow."

Global Managed WAN

Alvina has been looking to extend operations for Allofu into Asia, but the cost and complexity of the infrastructure alone are daunting. Shifting Allofu's manufacturing to Asia could really help reduce overall operational costs once it comes online, but cost reduction will also be a function of how quickly and seamlessly the extension is deployed. The board of directors may be getting closer to completing "The Big Deal," which sounds like a project to acquire a competitor, and this would add more complexity to the decision. Allofu already has operations in Europe and that deployment created its own set of challenges at the time. Additionally, the providers for Asia are such that several of the new locations that have been discussed are not easily accessed from the providers' backbones.

Rodney's workload is already overloaded and now he is hearing rumors of a potential trip to Asia. Rodney wonders if Alvina would throw a fit if she saw an expense report for a Kobe steak? Possibly a career-limiting dinner decision. Rodney was part of the team that helped deploy the European data center, so he has experience with multi-national networks. Rodney believes that if Allofu opens branches in Asia, the complications involved with making this happen would be far more complex than that of the European data center.

In an increasingly global economy, new challenges have naturally arisen. For organizations operating a global WAN, it is not uncommon to encounter situations whereby a given service provider lacks the required connectivity in certain countries within the carrier's network. This means that the different service providers have to agree to peer between one another in order to fulfill the organization's global connectivity needs, leading to frequently too costly and complex environments in which it can take a long time to even get a link between two sites.

VMware SD-WAN simplifies and optimizes connectivity, as it allows branch locations to connect to one another regardless of which service provider's connectivity (private circuits or internet) they are leveraging. Since VMware SD-WAN uses centralized orchestration to manage the entire network infrastructure of an organization, the result is a simpler, cheaper, and faster way to deploy and manage the WAN.

Alvina thinks the SD-WAN thing might genuinely make her life easier and it has been a while since anything about networking technology left her feeling this way. Alvina does not have time to deal with the reasons why she cannot do things - she needs more answers to her problems and SD-WAN is shaping up to answer several of her biggest challenges.
Rodney is primarily concerned with securing a business class upgrade for what appears to be several hours stuffed into a shiny metal tube hurtling through the sky. Rodney is worried that he really does not know where to begin with adding sites in Asia. But perhaps he doesn't need to travel to Asia after all, if all of this SD-WAN stuff is really true? That would be compelling, even though Rodney was sort of looking forward to that Kobe steak.

Enhanced Security in the Branch

Alvina wakes up on a regular basis with nightmares… or to the sound of the cat coughing up a hairball. The nightmares are recurring and always involve a breach of one form or another with her being escorted from the building with a cardboard box full of her belongings. Alvina goes to Hans the security architect (does that guy actually say anything other than "No?") to ask if SD-WAN is secure. Hans gives his usual answer "nothing is ever truly secure." Alvina leaves feeling that all hope is lost, but Hans did suggest several more security appliances he needed… what else is new. Alvina thinks that if SD-WAN can fundamentally change how her team approaches networking… then wonders "but what about security?". Alvina knows that continuing to purchase more networking appliances and shipping them to each of the branches is expensive and creates a never-ending refresh cycle. Why can't Allofu address its security issues the way it has solved its cloud issues?

Hans reminds Rodney of a character from his favorite Christmas movie, his last name even sounds a bit like "Gruber." Rodney was actually talking to a buddy during their goat yoga class, who mentioned that his security colleague had migrated their company to a cloud-based security offering as part of their SD-WAN implementation. Rodney figures that a cloud-based security solution would be simpler, but how does it handle traffic like SaaS that in some cases, is already secure? Can Allofu still use their existing security appliances if the security team decides to just find new ways to say "No?" Perhaps "Non," "No Way Jose"…"Nein"!

Network security is an ever-present challenge for all companies. Fully distributed security models of deploying physical security appliances at branch locations are expensive, complex to maintain and operate, and frequently lack application awareness. Another approach is to heavily centralize the security enforcement points in an effort to reduce costs and operational complexity, but this then means that all traffic has to be backhauled to this central location, which introduces another set of problems.

VMware SD-WAN solution addresses these issues in a number of ways. The first of these is the ability to deliver optimized access to cloud-based security services. This model uses third-party cloud-based security providers to which VMware SD-WAN can dynamically connect on a per-application basis. This reduces the requirement for physical appliances and greatly simplifies the deployment and management of the branch security services.

The ability to recognize and optimize SaaS applications is also built into VMware SD-WAN. This allows the network architect to designate different security policy for well-known SaaS applications than for non-trusted internet traffic.

Finally, to enable secure communication on the SD-WAN overlay, VMware SD-WAN supports standard IPsec encryption with built-in Certificate Authority (CA).

Alvina believes cloud-based security would further reduce the complexity of the stack of appliances Allofu uses today. Given the need to shift to the cloud, it is only natural that SD-WAN adoption would also support cloud security. Alvina figures that her SaaS users will get significantly better performance with SD-WAN and this will eliminate extra hops through the regional data center. Alvina feels strongly that SD-WAN is a solution that Rodney's team needs to vet as it appears to be the answer to many of her greatest challenges.
Rodney is on board with cloud security if it both works and silences Hans. Rodney will need to test and verify network performance with this solution, and to show Hans the improvements they can expect to achieve to ensure that he will sign off.

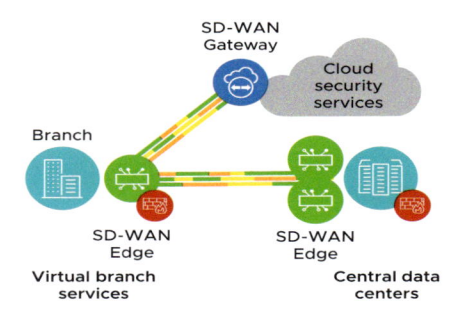

Figure 1.3: VMware SD-WAN Enhanced Security

The SD-WAN Vision: Network Edge

SD-WAN is revolutionizing the traditional wide area networking landscape: complex, hardware-intensive, and hub-and-spoke networks are transformed into cloud-friendly, cost-effective, and agile architectures. A flexible SD-WAN solution requires the following three components:

- VMware SD-WAN Orchestrators take care of simplified, UI-based configurations, and function as the management plane of the solutions. They can be hosted on different premises supporting different consumption models from fully managed to on-premises.

- VMware SD-WAN Gateways are the control plane of the solution, helping to steer traffic between edges and towards the most optimal location for the cloud onramp.

- VMware SD-WAN Edges take care of the data plane connectivity between users and where data is located (public cloud, data center, SaaS applications). Edges leverage the SD-WAN overlay to achieve connectivity to one another, making the SD-WAN data plane agnostic to the underlying transport technology.

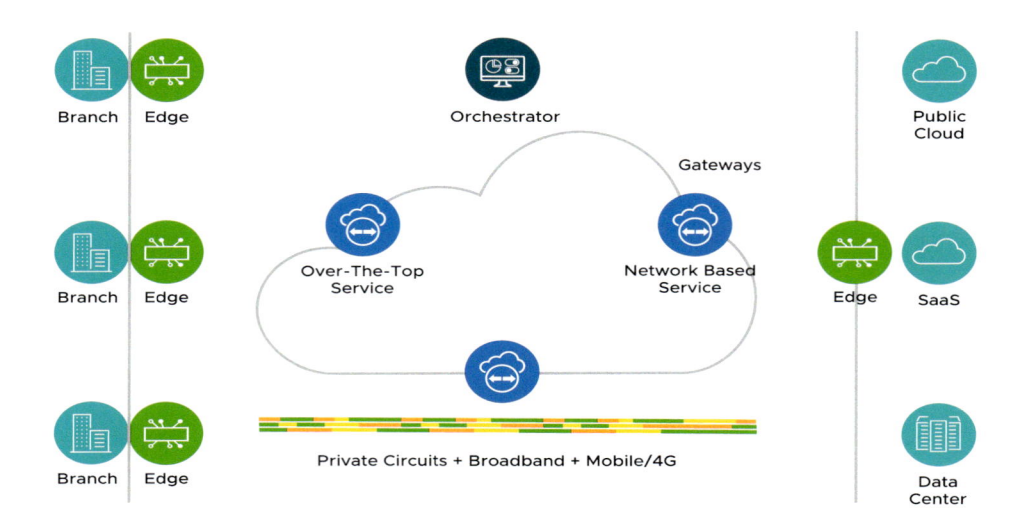

Figure 1.4: High-level SD-WAN architecture

Before diving into the SD-WAN vision, it is useful to clarify what is meant by the term **edge** in this context as this term means different things to different people. VMware is innovating a lot at the **edge**, namely in the realm of **device edge**, **compute edge**, and **network edge**.

Figure 1.5: VMware device edge, compute edge, and network edge

Device edge examples are VMware's Pulse™ and Workspace ONE product suites, which are optimized for IoT and end-user devices, respectively.

Compute edge refers to VMware's programs for compute (and other software-defined data center functionality such as storage and networking). VMware Cloud Foundation and its SaaS-based VMware Cloud on Dell EMC are examples of offerings in this space.

Network edge is the vision that enables the next wave of network transformation.

As business moves from data centers to **centers of data**, SD-WAN is expanding to absorb other functions such as compute, analytics, security, and multi-cloud that are critical to the enterprises where business is conducted, which is at the edge.

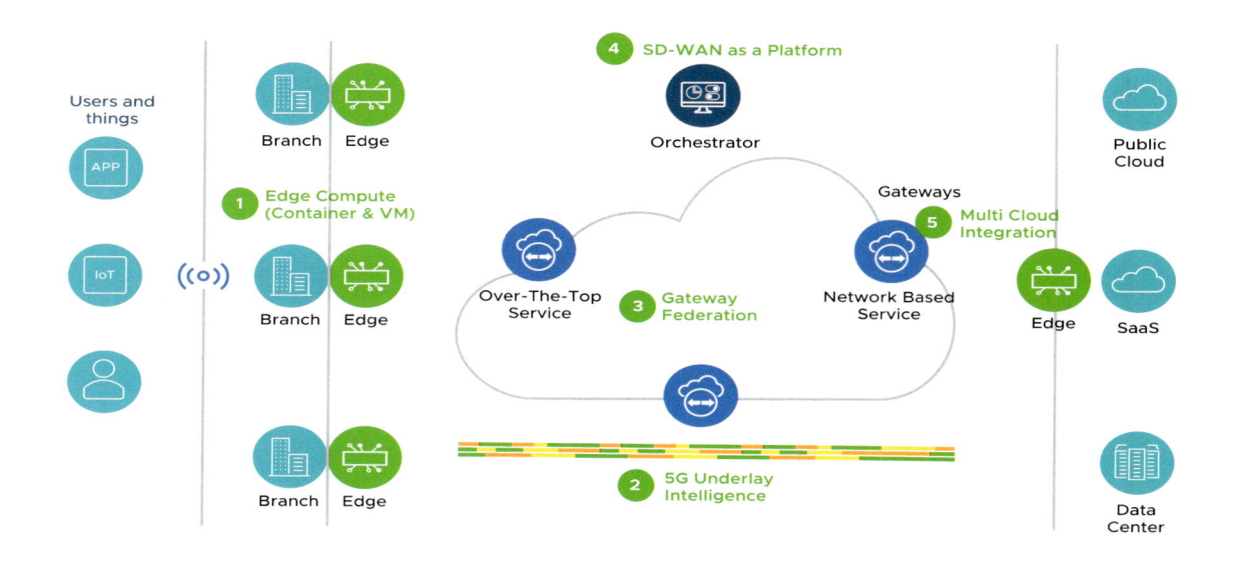

Figure 1.6: VMware vision for network edge

Network edge is evolving along five vectors beyond the core SD-WAN concepts.

1. Internet of Things (IoT) and mobility are indispensable aspects of business. Furthermore, there is an increased need to run distributed applications with application workloads at the edge. To support them, compute functionality is required inside the VMware SD-WAN Edge. This will allow workloads which have a low-latency requirement to be run at edge/branch/kiosk locations and will allow those workloads to offer services even while a branch or kiosk is disconnected from the network. Workloads can include virtual machines (VMs) as well as containers. The compute functionality will be incorporated from the same management plane where the rest of the application is running, i.e. in the data center.

2. In the telecom transport domain, VMware SD-WAN is combining the intelligence of its overlay SD-WAN with the intelligence and programmability of 5G. 5G network slicing is the capability of dividing the 5G network up into slices, with each slice having its own set of Service Level Agreements (SLAs) in terms of availability, bandwidth, latency, and jitter. Furthermore, these slices can be created in an automated/programmed fashion. This enables service providers to offer even more differentiated

SD-WAN services, as well as allowing enterprises to build better performing services.

3. With its unique gateway architecture, VMware SD-WAN is enabling control plane federation between VMware-hosted SD-WAN and those SD-WAN solutions hosted by its service provider partners. This capability will expand, enabling service provider partners to federate amongst each other, providing truly global coverage for all end-customers.

4. VMware SD-WAN will become the ubiquitous platform needed to support necessary business services such as security and others through network function virtualization (NFV), enabling a universal customer premise equipment (uCPE) approach. Rather than having to install and manage various pieces of equipment to offer the various network functions inside the branch, the uCPE approach allows a customer to have a platform where various third-party network functions can reside, providing huge simplification and cost saving. Furthermore, this will allow advanced analytics at the edge through Artificial Intelligence (AI), effectively creating dynamic self-healing and self-managing networks of the future.

5. VMware's multi-cloud strategy enables seamless workload placement, management, security, and networking across public cloud, private cloud, and compute edge locations. Multi-cloud is added to the holistic SD-WAN continuum, enabling the management of all network workloads from the edge to the cloud to the data center via pervasive business policies.

Organizations and their networking vendors are driven by the rise of IoT, 5G, AI, and multi-cloud, requiring a dramatic technological evolution in business requirements. VMware SD-WAN by VeloCloud continues to lead the market into the future, accelerating innovation at the network edge.

2. Components

After meeting to discuss the challenges with their network infrastructure, Alvina asks Rodney to come up with a new direction for the existing network and its future development. She is unsure if the network will be able to keep up with the future growth of the organization and its networking requirements such as availability, cloud connectivity, and increasing bandwidth demand.

Alvina's priority now becomes Rodney's priority and he cancels his goat yoga plans for the evening. He begins to explore the options available that will enable Allofu to address those issues and support its business growth. During his research, he comes across an online video where Steve Woo at Tech Field Day gives a brief technical overview of VMware SD-WAN. Rodney knows VMware, as&nbps;his infrastructure colleagues have been running VMware server virtualization for&nbps;years. Rodney wasn't aware, though, that VMware had an offering for SD-WAN! Eager to find out more, he reaches out to Leia, his VMware System Engineer, and sets up a meeting with her to discuss the VMware SD-WAN solution. The following chapters cover his research journey.

Solution Components

The VMware SD-WAN solution is comprised of three components that provide organizations with an optimized platform to deliver high-performance, reliable branch access to cloud services, private data centers, and SaaS-based enterprise applications. These components are all provided via a cloud-as-a-service delivery model, substantially lowering time to market and total operating costs. Each of the VMWare SD-WAN solution components are listed below and detailed in this chapter.

1. VMware SD-WAN Orchestrator
2. VMware SD-WAN Gateway
3. VMware SD-WAN Edge

VMware SD-WAN Orchestrator by VeloCloud

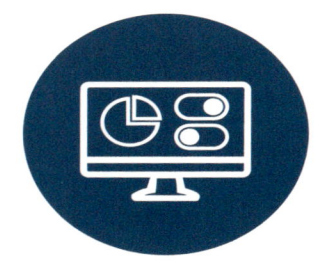

Figure 2.1: VMware SD-WAN Orchestrator

The **VMware SD-WAN Orchestrator** is a cloud-delivered, multi-tenant portal providing centralized management, configuration and monitoring. The orchestrator provides the ability to deliver business-driven policy abstraction, enabling rapid deployments and zero-touch operations. Additionally, it exposes a rich set of APIs that can be utilized to provide management, troubleshooting, and Operations Support System/Business Support System (OSS/BSS) integration. The orchestrator is hosted by VMware in Statements on Standards for Attestation Engagements (SSAE) 16 SOC 2 Type II data centers providing a fully redundant, highly available management portal. Some of its functions are listed below:

- Configuration includes items such as edge activation, provisioning, and business policies. It is the key element for zero-touch provisioning of new VMware SD-WAN Edges.

- Monitoring includes providing user information about link and flow statistics as well as link quality scoring. It also covers application-related statistics. This can be used to provide analytics about business compliance for customers or network engineers.

- Diagnostic support includes providing users a centralized interface for accessing features such as remote diagnostics, packet captures or other relevant support tools for troubleshooting and operating the solution.

VMware SD-WAN Gateway by VeloCloud

Figure 2.2: VMware SD-WAN Gateway

The **VMware SD-WAN Gateway** is also cloud-delivered and multi-tenant. The gateways are deployed by VMware and its partners at top-tier network points of presence and cloud data centers around the world, facilitating the full range of VMware SD-WAN benefits. VMware SD-WAN Gateways provide a scalable and distributed infrastructure with the advantages of hosted, network-as-a-service flexibility. They also provide the ideal architecture for optimized access to cloud applications and data centers, as well as access to private network backbones and traditional enterprise sites.

The VMware SD-WAN Gateway can function in two independent roles, each providing separation of services. These services are referred to as planes and the gateway can provide both control plane services and data plane services. If needed, both can be combined to provide a cohesive platform that is used to rapidly and efficiently scale WAN services. When the gateway is only participating as a control plane element, its role is referred to as the **VMware SD-WAN Controller**. The control plane provides a number of important functions including:

- Discovery of link bandwidth and IP addresses.
- Route information distribution and updates.

In addition to functioning as a controller, it can also provide data plane services for end-user applications. When the VMware SD-WAN Gateway's data plane capability is enabled, its role serves as a multi-tenant SD-WAN endpoint for SaaS onramp and service provider integration and is referred to as a **VMware SD-WAN Gateway**. VMware SD-WAN Gateways are stateless elements and designed to be highly scalable, allowing for quick expansion and recovery.

VMware SD-WAN Edge by VeloCloud

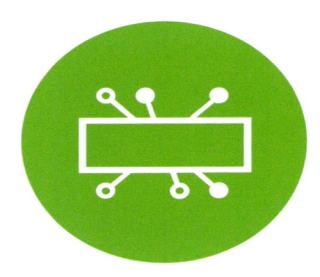

Figure 2.3: VMware SD-WAN Edge

The **VMware SD-WAN Edges** are available as easy-to-install appliances for remote branches and data centers with a range of throughput, interfaces, integrated wireless and LTE connectivity options. Dynamic routing enables policy-based overlay insertion for both in-path and out-of-path deployments. High availability deployments are also supported. In addition to the physical appliance options, the

VMware SD-WAN Edge is available as a virtualized network function (VNF) for software-only deployments on standard x86 servers including virtual CPE devices. The VMware SD-WAN Edge receives all policies and configurations from the VMware SD-WAN Orchestrator, classifies traffic using a deep-packet application recognition (DAR) engine and applies policies based on real-time link quality. Traffic steering decisions are made locally on the VMware SD-WAN Edge. VMware SD-WAN Edges are deployed in branch locations, data centers, or IaaS platforms such as Amazon AWS or Microsoft Azure.

> **VMware SD-WAN Overlay**
> VMware SD-WAN utilizes the VeloCloud Multi-path Protocol (VCMP) to enable IPSec-secured transport of data and control information over any kind of transport. VCMP tunnels are established between two VMware SD-WAN Edges or between a VMware SD-WAN Edge and a VMware SD-WAN Gateway. VMware SD-WAN overlay is comprised of VCMP tunnels which provide an abstraction layer from the underlying transport. VCMP uses IANA-registered UDP port 2426.

Form Factors

VMware SD-WAN Orchestrators and VMware SD-WAN Gateways are deployed as software-only entities. VMware SD-WAN Edges can be deployed as virtual (software) or as hardware appliances.

The hosted VMware SD-WAN Orchestrator and VMware SD-WAN Gateways are usually deployed in a virtual form in either public clouds such as AWS, Azure, or through large ISP data centers for optimal access for the users. As of this writing, software images are available either as an OVA or QCOW2.

VMware SD-WAN Edges

Figure 2.4: VMware SD-WAN Edge form factors

Physical VMware SD-WAN Edge

For the VMware SD-WAN Edge, there are several different physical appliances available. The primary differences include the bandwidth/throughput performance, number of tunnels, and number as well as types of interfaces in order to support small branch sites, all the way up to data center requirements.

Virtual VMware SD-WAN Edge

VMware SD-WAN Edges can be deployed as VNFs on a uCPE or on any kind of x86 hardware. This flexibility allows for virtualization with either VMware vSphere or KVM.

Cloud-based VMware SD-WAN Edge

VMware SD-WAN also provides virtual images for VMware SD-WAN Edges through AWS and Azure marketplaces. This option allows for the termination of customer networks in virtual infrastructures hosted on the public cloud infrastructure in order to extend their networks into the cloud.

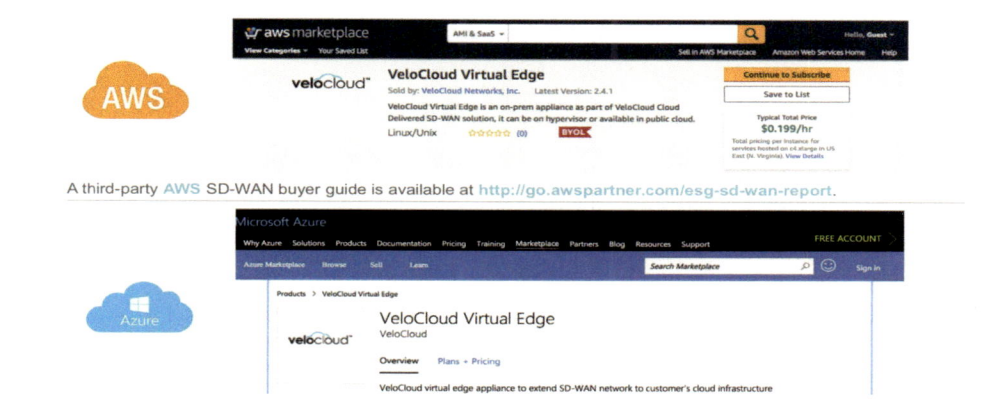

Figure 2.5: SD-WAN at AWS and Azure marketplaces

Placement of Components

VMware SD-WAN Orchestrator

The most common deployment option for the VMware SD-WAN Orchestrator is cloud-delivered, which means it is operated and maintained by VMware. VMware takes care of all infrastructure requirements such as needed storage, compute and databases. The user of the VMware SD-WAN Orchestrator can be a service provider partner or an end-customer.

The cloud-delivered VMware SD-WAN Orchestrators are distributed all over the world across three regions (Americas, Asia, and Europe), which allows for disaster recovery enabled by default, utilizing SSAE16 SOC 2 Type II data centers. This provides a 99.99% availability SLA per month.

The VMware SD-WAN Orchestrator can also be placed on-premises by service providers or organizations.

VMware SD-WAN Edge

VMware SD-WAN Edge is commonly placed at the edge of the network, either in branches or data centers. It can also be deployed in IaaS such as Amazon Web Services (AWS) or Microsoft Azure.

VMware SD-WAN Gateway

VMware SD-WAN Gateway can be deployed in two options:

- When the VMware SD-WAN Gateway is deployed as a **cloud-hosted gateway,** it provides optimized and secured access to SaaS applications for branch sites. These gateways are hosted by VMware and are geographically distributed all over the world.

- When the VMware SD-WAN Gateway is deployed as a **partner gateway,** it is on-premises and managed in a service provider network. Partner gateway is deployed in a dual-arm fashion, with one network interface connecting to the internet and the other network interface connecting to a service provider's MPLS backbone. It provides access to service provider infrastructure and services.

Communication between Components

Management Plane

The management plane is the communication channel between:

- VMware SD-WAN Edges and VMware SD-WAN Orchestrator
- VMware SD-WAN Gateways and VMware SD-WAN Orchestrator

VMware SD-WAN Edges use the management plane to receive configurations and policy updates, as well as to upload statistics and event logs. The management plane traffic is carried between components via TLS 1.2. Management plane traffic is first initiated by the VMware SD-WAN Edge or VMware SD-WAN Gateway towards the VMware SD-WAN Orchestrator upon initial activation.

For the management of the VMware SD-WAN Edge and VMware SD-WAN Gateway, heartbeats are sent towards the VMware SD-WAN Orchestrator every 30 seconds. In the event that four consecutive heartbeat packets are lost, the VMware SD-WAN Edge or VMware SD-WAN Gateway is declared offline. The VMware SD-WAN Orchestrator uses the heartbeat response to carry any configuration updates and requests for software updates of VMware SD-WAN Edges. Flow and link status statistics are also sent every five minutes by the VMware SD-WAN Edge to the SD-WAN Orchestrator.

Control Plane

The control plane is the communication channel between VMware SD-WAN Edges and VMware SD-WAN Gateways. The function within the VMware SD-WAN Gateway that facilitates communication with the VMware SD-WAN Edges is referred to as the VMware SD-WAN Controller.

The control plane is used for the exchange of routing information, and the detection of WAN characteristics including IP addresses and bandwidth. VMware SD-WAN Gateways serve as SD-WAN route reflectors, and facilitate route exchanges and updates among VMware SD-WAN Edges. Routing table is then stored locally on VMware SD-WAN Edges for forwarding decisions.

The control plane leverages the VCMP tunnel which is established from the VMware SD-WAN Edges to the assigned VMware SD-WAN Gateways. The control plane communications are also encrypted using IPsec.

Data Plane

The data plane is the communication channel between two VMware SD-WAN Edges or between a VMware SD-WAN Edge and a VMware SD-WAN Gateway. Data plane traffic is encapsulated using VCMP and secured using IPsec encryption.

> Due to the separation of management, control and data planes, VMware SD-WAN Edge branch-to-branch traffic will continue to flow, based on the local routing table entries, in the event the VMware SD-WAN Edge loses connectivity to the management or control plane. There is no impact on packet forwarding for established SD-WAN peers and prefixes. The operator only loses the ability to make configuration changes on the VMware SD-WAN Edges until the management or control plane connectivity is re-established.

Figure 2.6 shows the management, control, and data plane communications between VMware SD-WAN components.

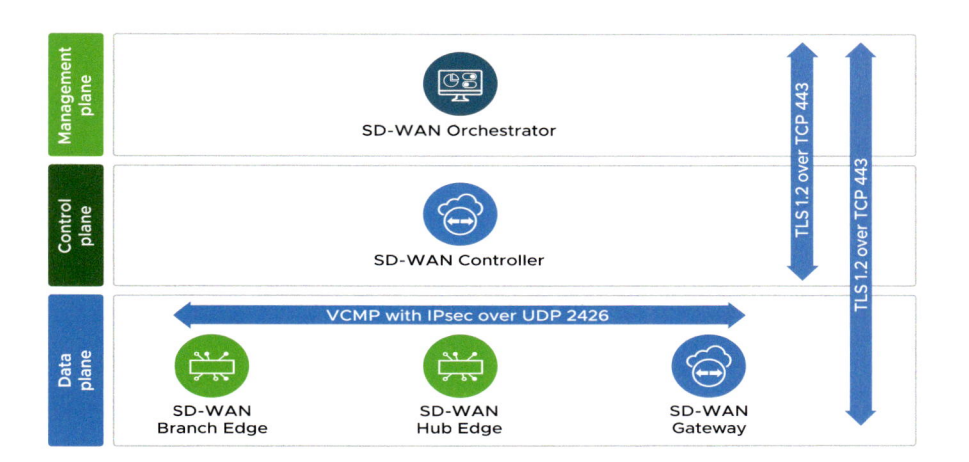

Figure 2.6: Management, control, and data plane communications between VMware SD-WAN components

Components 37

After meeting with Leia, Rodney has gained a good understanding of VMware's SD-WAN and its components. Rodney sees promise and it sure sounds as though SD-WAN theoretically could help resolve some if not all of his primary pain points. Alvina sat in on the first part of the meeting as well and she clearly was enthusiastic about the business benefits which were discussed. Rodney has seen more than his fair share of "marketectures" and is not holding his breath. Rodney has deployed more than one "revolutionary" product after another and has more battle scars than he cares to recall. Rodney knows that he needs to temper Alvina's enthusiasm with a solid dose of reality to avoid gathering any more scars. Rodney needs to keep digging deeper and get past the slideware and into where the rubber meets the road.

3. Getting Started

Alvina checks in with Rodney on how far along he is on his SD-WAN research. Alvina is anxious to see the outcomes of this project and knows that she needs to keep this herd of cats moving. One thing Alvina has learned about her team is that they can easily get caught up in an endless R&D cycle and she has to keep them focused on the business outcome!

Rodney explains that he now has a good understanding of all of the VMware SD-WAN components and that he sees potential in the VMware SD-WAN components, but he still needs to wrap his mind around a number of important concepts. Rodney needs to get his hands on the VMware Orchestrator so that he can comprehend how he would actually activate branches and apply the necessary routing and business policies. Rodney doesn't really know what a "business policy" even is at this point! Sounds like marketing fluff, let's see this zero-touch-provisioning magic first! Rodney schedules a meeting with Leia to walk him through the branch activation process as he figures it's the next logical step to take. Rodney continues to temper Alvina's enthusiasm or things could well get worse before they get better—he has seen that movie before!

High Level VMware SD-WAN Deployment

One of the advantages of VMware SD-WAN is the simplicity that the solution offers in regards to provisioning and site bring-up. In order to understand this further, a compare and contrast is needed between a traditional WAN edge deployment and a VMware SD-WAN Edge deployment. Figure 3.1 depicts an enterprise deployment with two data centers and four branches.

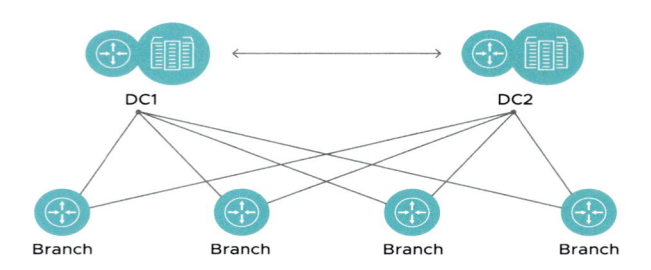

Figure 3.1: Example of enterprise deployment

Figure 3.2 depicts a typical workflow for bringing up a site in a traditional WAN environment. One or more network engineers must decide on connectivity options, then configure the WAN routers via a Command Line Interface (CLI) to enable various features such as routing, Access Control Lists (ACLs), IP addressing, Quality of Service (QoS), Virtual Routing and Forwarding (VRFs), and High Availability (HA) amongst others. Often this is done in a staging area and once the devices are configured, the next step is to ship the WAN router to the branch. If there are any mistakes made during the manual configuration process, the deployment of the traditional WAN edge will not be successful.

This approach is time-consuming, error-prone, and may require multi-vendor expertise. Most importantly, the process is static in nature and is painstakingly hard to ensure configuration continuity across the organization.

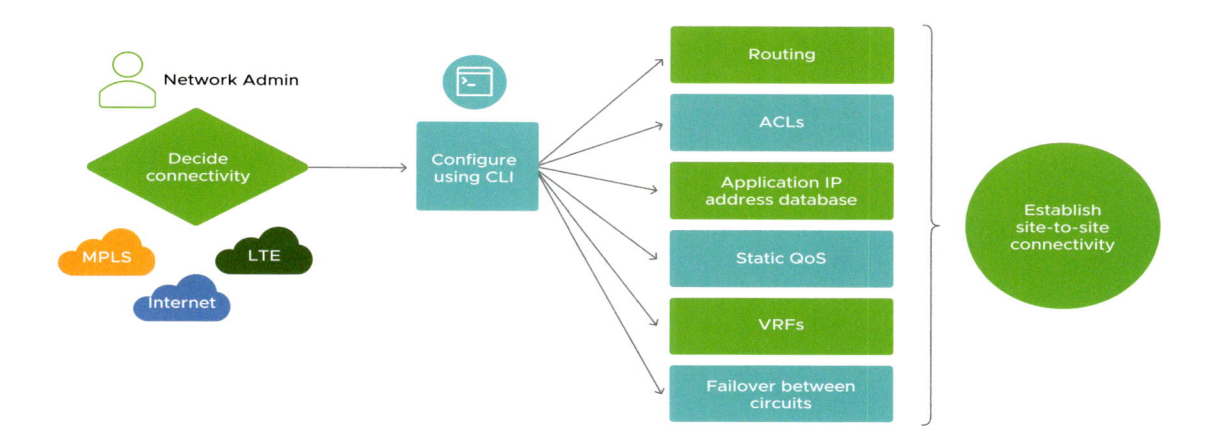

Figure 3.2: Traditional deployment workflow

In contrast to the complexity associated with a traditional WAN, VMware SD-WAN simplifies an organization's WAN management via a policy-driven user interface (UI) that allows them to easily translate business objectives into network policy. This simplification is achieved using three steps, all delivered in a Cloud-as-a-Service orchestration model.

Figure 3.3: VMware SD-WAN deployment workflow

In the following, each of these steps will be discussed in more detail.

Step 1: Profile Creation

Getting Started 41

Profiles define standard configuration elements applied to one or more VMware SD-WAN Edges at the time of provisioning. A profile is a named configuration that defines the following:

- Cloud VPN settings
- Routing protocols
- Business policies such as application-aware QoS
- Security
- Etc

Figure 3.4 provides an example of multiple `profiles` that can be used for sites with similar properties. In this case, both a `branch profile` and a `data center` profile are used by the customer.

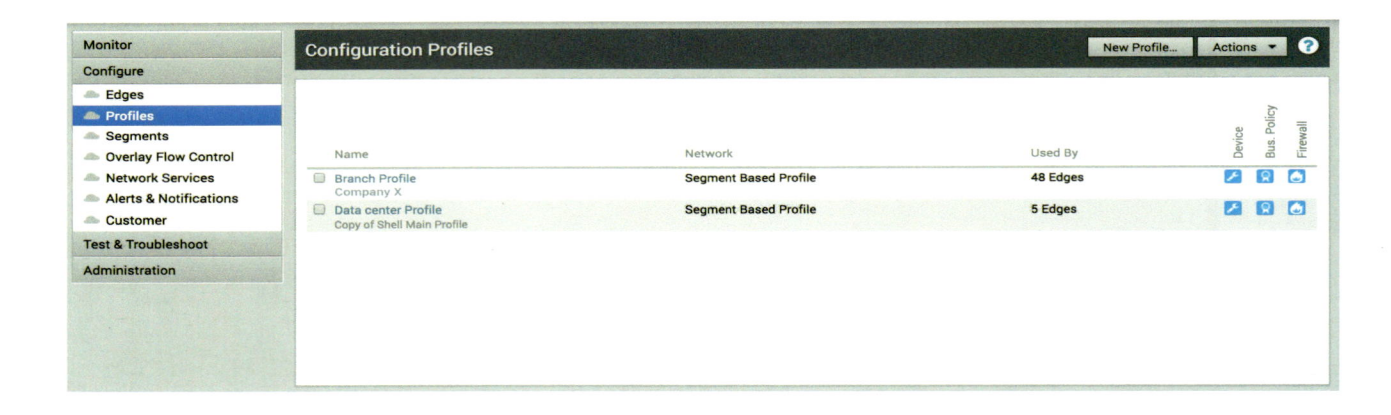

Figure 3.4: Profile configuration

Step 2: Business Policy

VMware SD-WAN's business policy framework provides out-of-the box smart defaults settings which specify the traffic behavior and Quality of Service for over

3,000 applications. Business policy settings work in close conjunction with VMware's Dynamic Multi-Path Optimization (DMPO). DMPO will be covered in section *Application Performance, Routing and Cloud Access*.

Figure 3.5 shows a screenshot of the VMware SD-WAN `Business Policy` smart defaults.

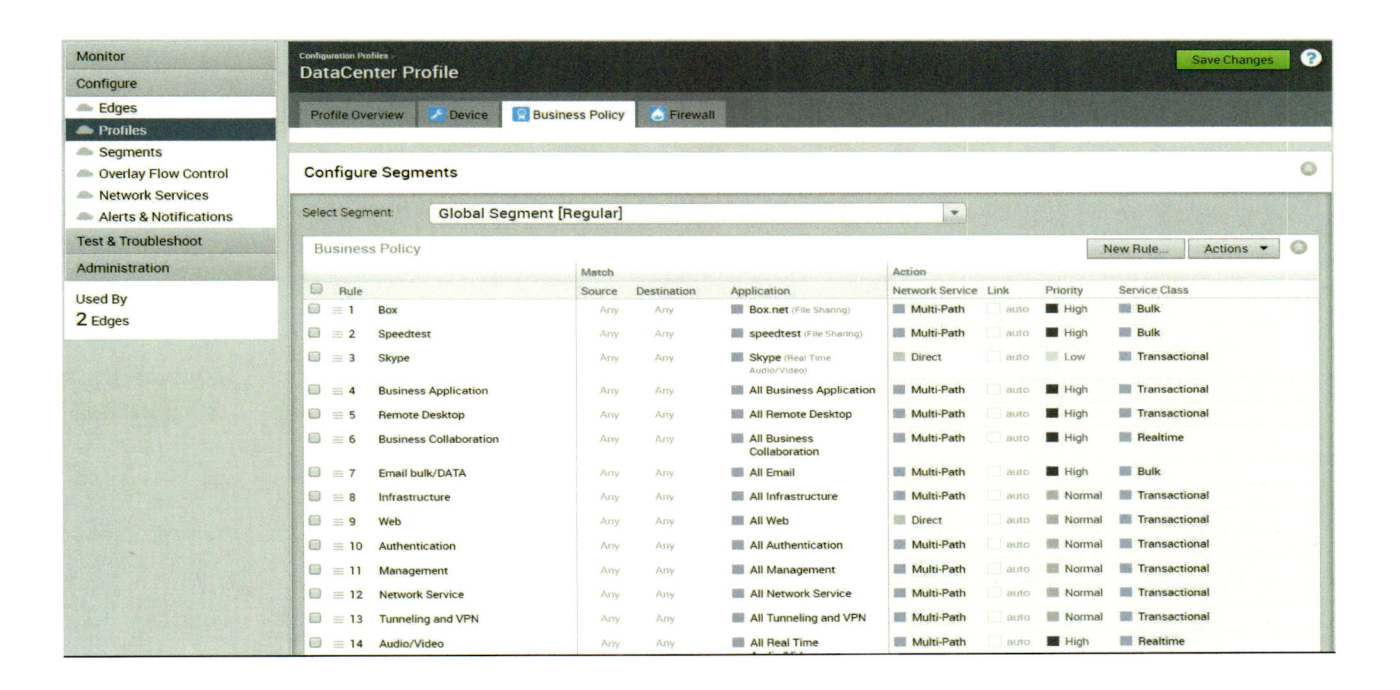

Figure 3.5: VMware SD-WAN Orchestrator UI view of Business Policy

Step 3: VMware SD-WAN Edge Provisioning

Provisioning an edge is as simple as creating a new edge and activating it via a zero-touch provisioning model. The next chapter will provide more detail on the two methods of activating a VMware SD-WAN Edge. Figure 3.6 is a screenshot which highlights the process of `provisioning a new edge` within the VMware SD-WAN Orchestrator.

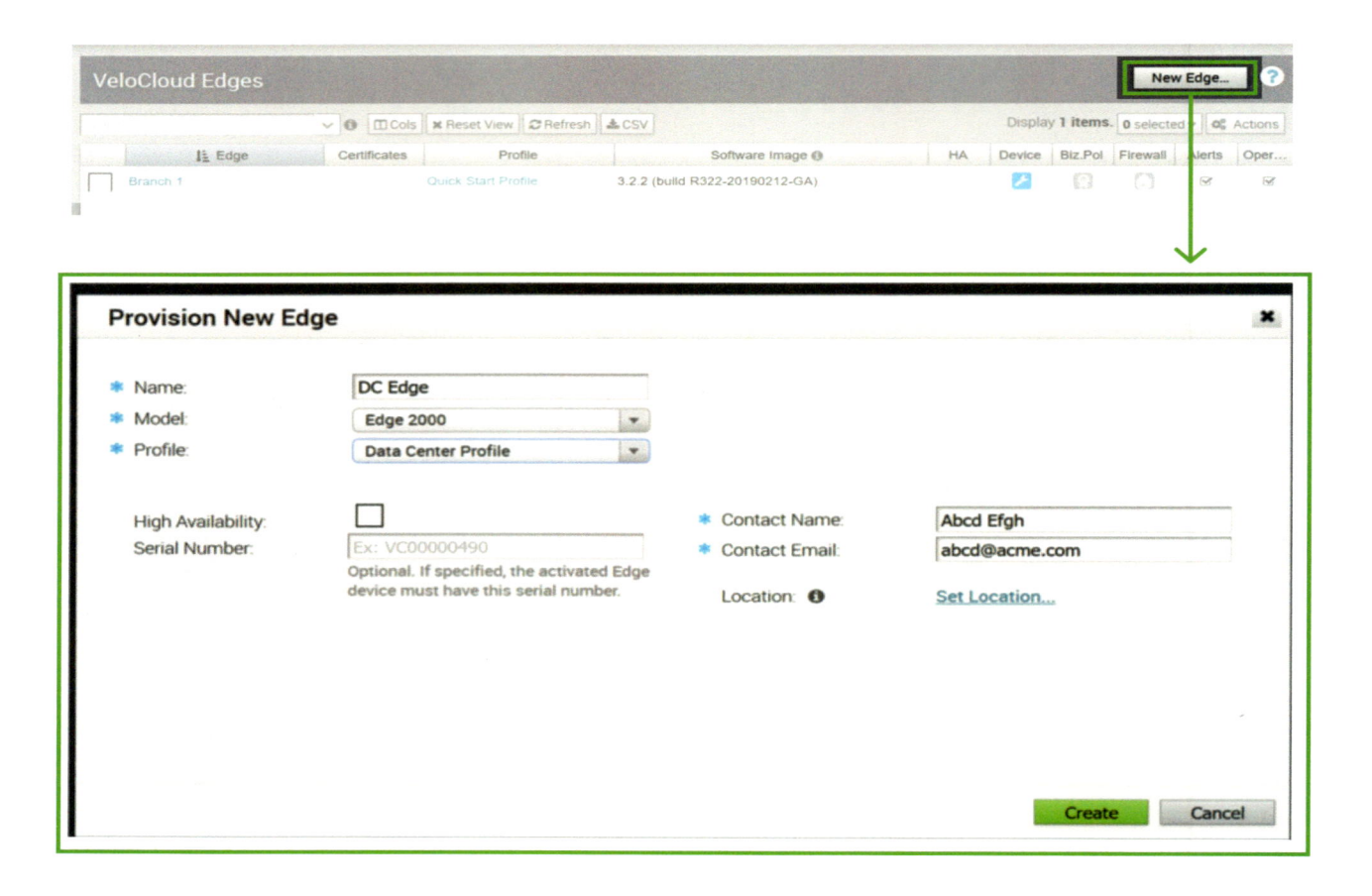

Figure 3.6: VMware SD-WAN Edge creation

Key Takeaways:

1. VMware SD-WAN simplifies provisioning of branch locations.
2. Profiles can be used to deploy multiple VMware SD-WAN Edges with common attributes while enforcing enterprise compliance.
3. Business policy contains out-of-the-box smart defaults that specify the default traffic behavior and QoS of more than 3,000 applications.
4. Provisioning an edge is as simple as creating a new edge and activating it.

Zero-Touch Provisioning

VMware SD-WAN Edge Activation

In a traditional WAN router deployment, each router needs to be individually configured by a network engineer through CLI. This task is time-consuming and error prone. To make the deployment simple and scalable, VMware SD-WAN offers zero-touch provisioning to facilitate the deployment of the VMware SD-WAN Edge. The goal of the activation is to allow a VMware SD-WAN Edge to be registered to the VMware SD-WAN Orchestrator such that any further device operations are centrally managed by the VMware SD-WAN Orchestrator. The VMware SD-WAN Orchestrator does not require the engineer to login to the VMware SD-WAN Edge individually for configuration and monitoring. In the VMware SD-WAN solution, the VMware SD-WAN Orchestrator pushes the appropriate configurations to the VMware SD-WAN Edge directly.

There are currently two types of activation supported—pull activation and push activation.

Pull Activation

The pull activation delivers the following benefits:

1. No pre-staging required.
2. No security risk if an appliance is lost.

Figure 3.7 illustrates pull activation:

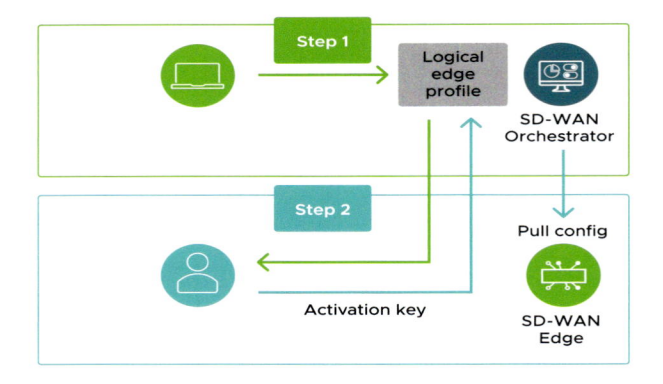

Figure 3.7: Pull Activation

Steps for pull activation:

1. The network engineer provisions a new VMware SD-WAN Edge in the VMware SD-WAN Orchestrator and sends to the onsite installer an auto-generated activation URL via an email. See Figures 3.8 and Figure 3.9 below. Please note the onsite installer does not require networking expertise.

2. Once the edge arrives at the customer site and is attached to the network, the onsite installer connects via Wi-Fi or Ethernet to the LAN and clicks on the activation URL.

> Upon activation initiation, the VMware SD-WAN Edge contacts the VMware SD-WAN Orchestrator and downloads any necessary software update and configurations.

In instances where the WAN circuit is using dynamic IP addresses, the activation URL contains only the activation key and the VMware SD-WAN Orchestrator FQDN. Figure 3.8 is a screenshot of such example.

46   SD-WAN 1:1

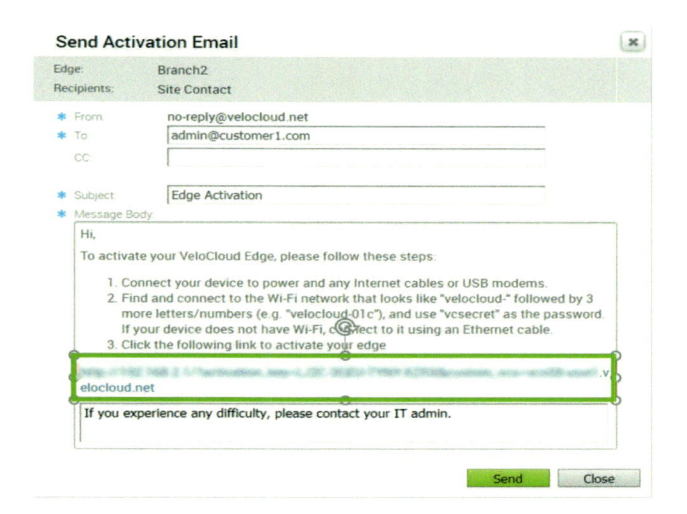

Figure 3.8: Activation URL when WAN circuit is with dynamic IP

In instances where the WAN circuit is using static IP addresses, the activation URL includes an additional field called **configuration**. The VMware SD-WAN Edge intercepts the URL activation request and provisions the static interface IP address and default gateway per instructions encoded in the URL. Figure 3.9 is a screenshot of such an example.

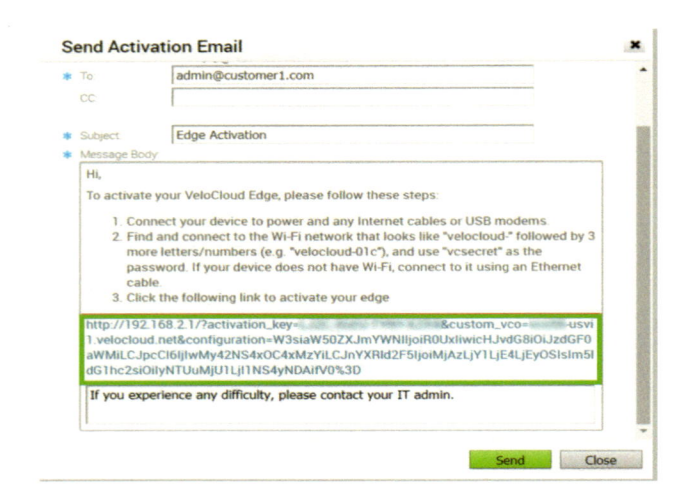

Figure 3.9: Activation URL when the WAN circuit has a static IP address

> **TECH TIP**
>
> In instances where public and private WAN circuits are used, the public WAN should be connected to the lower numbered GE interface.

Push Activation

Push activation has the same benefits of pull activation, but requires no onsite user interaction.

Figure 3.10 illustrates the high-level workflow of push activation.

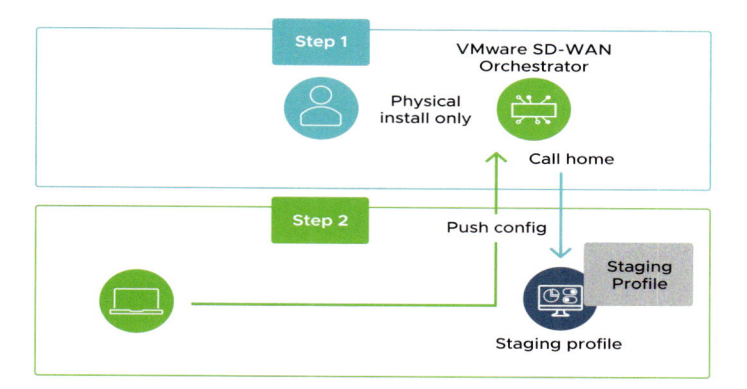

Figure 3.10: Push activation

Steps for push activation:

1. Once the VMware SD-WAN Edge is connected to the network, it performs a call home to a cloud-based Redirector, which assigns the edge to a VMware SD-WAN Orchestrator based on the edge's serial number.

2. The edge is then created and activated in a staging profile on the VMware SD-WAN Orchestrator. The network engineer can reassign the edge to the desired profile.

Key Takeaways:

- VMware SD-WAN simplifies rollout of new sites by Zero-Touch Provisioning.
- There are two models for VMware SD-WAN Edge activation: push and pull.
- With pull activation, the person on site just needs to cable and power the VMware SD-WAN Edge and click a link in an activation email.
- With push activation, the person on site just needs to cable and power the VMware SD-WAN Edge.

Getting Started

After Leia's office visit and an excellent lunch—who knew that what looked scarily like a pile of weeds from the garden might actually taste good? It just goes to show that with ranch dressing and some bacon bits, even a pile of weeds isn't so bad—Rodney's confidence is rising. This SD-WAN project is starting to look real and he was able to walk through the activation process with Leia and Raycliff, the VMware SD-WAN specialist SE, using the demo appliance the team brought along to the meeting. The process was surprisingly quick and simple. Rodney even made sure to go through the process a couple of times to make sure that he really understood it—surely he was missing something as it felt almost too simple! Rodney realizes that everything looks awesome in a lab. If you really want to understand how something works, put it into production, and put some users on it-they will surely come up with new ways to break your best-laid plans! Rodney asks the VMware team if Allofu should run a pilot to see how the solution works in their network. To Rodney's surprise, the VMware SD-WAN specialist SE suggests they run a pilot on their most problematic branch instead. Rodney and his team all look at each other and say in unison "North Dakota!"

Rodney asks Alvina to step into the meeting to bounce the idea of running a VMware SD-WAN pilot at the North Dakota branch. Alvina visibly flinches at the mere mention of the North Dakota branch, as she has just wrapped up a call with Bob the branch manager from North Dakota. Bob has a different outlook on life that appears to be "No good news is news"! The guy is never happy! Alvina realizes that things really can't get worse at the North Dakota branch, so why not? If it works, it's a win. If it doesn't work then it's status quo. Alvina approves and gives her standard "Make it so Number One" directive.

Rodney still needs to figure out how to architect the actual deployment before he can start inserting a new appliance onto his network. It needs to be transparent to the users, and most importantly, it needs to improve the performance of the applications and the never-ending stream of trouble tickets opened by Bob!

4. Topologies

Now that Rodney has the beginnings of a plan to run a pilot at the North Dakota branch, he needs to get down and dirty with the execution plan. As a network architect, Rodney loves the design phase of his job the most. Putting the puzzle pieces together is where this really gets real! Rodney invites the VMware SD-WAN specialist SE back in for a good old-fashioned whiteboard session to explore the various topologies and how to design for scalability and high availability. Raycliff, the VMware SD-WAN specialist SE, agrees to come in and conduct a workshop to help Rodney and his team understand their options. Alvina checks in with Rodney on his progress and reminds him that this is not a science project, that speed is one of the outcomes, so how about moving this project along a bit faster? Rodney is careful to not roll his eyes but tells Alvina he is doing the best that he can. He goes back to researching the 10 best steak restaurants in America.

The next few chapters cover specific in-depth topics, backed up by screenshots from the orchestrator or detailed diagrams and topologies.

Chapter 4 covers *Overlay Options* and their configurations. Different overlay types are needed to connect to private (MPLS) vs public (internet) networks. Using these overlays, the various topologies are created as explained in the *Overlay Topologies* chapter. The different hosting options available for the SD-WAN components are explained in the *Hosting Variants* chapter. *Site Topology and Redundancy* are subsequently covered.

Overlay Options

The VMware SD-WAN Edge uses interfaces that can be used to simultaneously route to the underlay and establish SD-WAN Overlay. A single routed interface can also have multiple public and/or private overlays allocated to it, separated by 802.1q VLAN tags.

> The underlay is the existing network infrastructure and can be either public (internet) or private (MPLS).
> An overlay is built on top of an underlay utilizing VCMP tunnels.

Public and Private Overlays

A public WAN overlay is defined as one that runs over a public underlay network where a VMware SD-WAN Gateway is reachable. A public WAN overlay is automatically detected and created on each edge when an interface to a public network comes up, and after a successful negotiation with a VMware SD-WAN Gateway.

A SD-WAN overlay can be configured to be auto-detected or user-defined. By default, all routed interfaces are configured to auto-detect the overlay on that interface when a cable is inserted. As seen in Figure 4.1, this is done by sending out a "tunnel initiation message" to a list of assigned VMware SD-WAN Gateways.

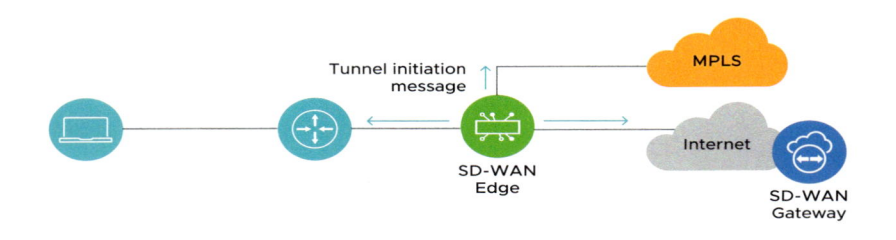

Figure 4.1: VMware SD-WAN Edge tunnel setup

A valid response from a VMware SD-WAN Gateway and subsequent successful bandwidth test (or a manual definition of available bandwidth) will result in the setup of VCMP tunnels from the VMware SD-WAN Edge to one or more VMware SD-WAN Gateways and this newly discovered WAN overlay is reported to the VMware SD-WAN Orchestrator. This behavior is determined by the `Auto-Detect Overlay` menu option under `Configure > Edges > {Edge} > Device tab > {Interface_number}` (see Figure 4.2). The Public IP of the overlay is derived from the tunnel information exchanged with the VMware SD-WAN Gateway.

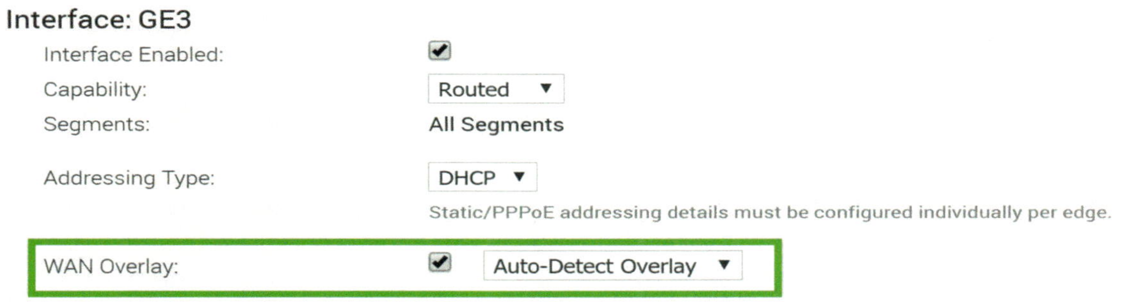

Interface: GE3

Interface Enabled:	✔
Capability:	Routed ▼
Segments:	All Segments
Addressing Type:	DHCP ▼
	Static/PPPoE addressing details must be configured individually per edge.
WAN Overlay:	✔ Auto-Detect Overlay ▼

Figure 4.2: A routed interface on an VMware SD-WAN Edge with the default WAN overlay setting applied

Conversely, a private WAN overlay is user-defined and is carried over a private network where a VMware SD-WAN Gateway is not reachable. The IP address of the interface associated with a user-defined private overlay is used to populate the peer table of other VMware SD-WAN Edges in the enterprise. This provides the necessary tunnel destination information required to initiate and maintain tunnels on this private underlay network.

For interfaces attached to a private underlay, the setting for `WAN Overlay` in `Configure > Edge > {edge name} > Device tab > {interface}` must be set to `user-defined Overlay`. This will instruct the edge to pass the interface IP address associated with the user-defined overlay, aka private address, to the VMware SD-WAN Orchestrator. The VMware SD-WAN Orchestrator will in turn pass this private address to other VMware SD-WAN Edges that are configured for a user-defined overlay, so those remote VMware SD-WAN Edges can learn the tunnel endpoint IP address over the private network, needed when building an SD-WAN overlay. Figure 4.3 shows an example of an interface configured for a `user-defined overlay`.

Topologies

Interface: GE2	
Interface Enabled:	☑
Capability:	Routed ▾
Segments:	All Segments
Addressing Type:	DHCP ▾
IP Address:	n.a
CIDR prefix:	n.a
Gateway:	n.a
WAN Overlay:	☑ **User Defined Overlay ▾**

Figure 4.3: VMware SD-WAN Edge interface connected to a private network with user-defined overlay selected

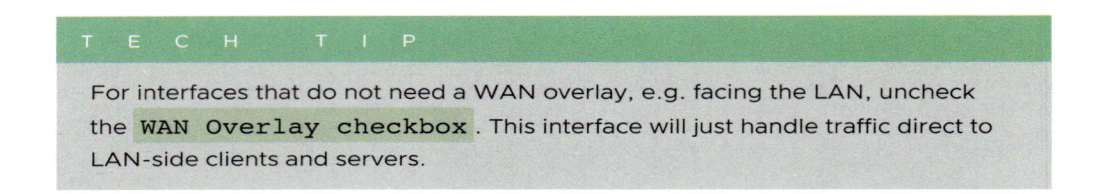

> **TECH TIP**
>
> For interfaces that do not need a WAN overlay, e.g. facing the LAN, uncheck the `WAN Overlay checkbox`. This interface will just handle traffic direct to LAN-side clients and servers.

Once an interface is configured for a user-defined WAN Overlay, you then simply create a new user-defined `WAN Overlay` and bind it to the interface already configured for user-defined overlay as per above. For example, in Figure 4.4 we show the `user-defined WAN Overlay` configuration available under `Configure > Edge > {edge name} > Device tab >` Add `user-defined WAN Overlay` and `Link Type` is set to `Private`.

Figure 4.4: New user-defined WAN overlay with a link type of private, being bound to GE2 which has already been set to carry a user-defined overlay

There is an optional configuration within the `user-defined WAN overlay configuration` to define the IP address, next-hop IP, and VLAN ID, otherwise it will inherit those properties directly from the interface.

> **TECH TIP**
>
> Always set the WAN overlay to `user-defined` for private networks. If the WAN overlay is left as `auto-detect overlay,` and a hosted VMware SD-WAN Gateway is reachable via the MPLS network (typically via a data center firewall), the VMware SD-WAN Edge will incorrectly create a WAN overlay of type "public" via a private network, and, in this case, the public address of the firewall in the data center will likely become the tunnel endpoint address for this VMware SD-WAN Edge, which is not desirable.

SD-WAN Service Reachable

Under private overlays, there is an option to enable a feature known as `SD-WAN Service Reachable`. Selecting this option informs the VMware SD-WAN Edge that there exists on the private network a means to breakout to the cloud-hosted VMware SD-WAN Gateways and Orchestrator. When selecting this option, a list of public /32 addresses will be shown that represents the cloud-hosted VMware SD-WAN Gateways and Orchestrator. A network engineer should advertise those prefixes into the private network, in case advertising a default route into the private network is not desirable.

Topologies

> **T E C H   T I P**
>
> Always enable the `SD-WAN Service Reachable` setting when the private network provides reachability to the internet, as this feature provides an SD-WAN Overlay failover path to cloud-hosted VMware SD-WAN Gateways and Orchestrator in the case where the links to the public internet fail.

Setting a Private Network Name

In some cases, VMware SD-WAN Edge in the data center may be the aggregation point for multiple MPLS provider networks e.g. MPLS A and MPLS B. In this scenario, all VMware SD-WAN Edges connected to MPLS A should not attempt to build tunnels to the private IP addresses of VMware SD-WAN Edges connected to MPLS B. To help with this, the VMware SD-WAN solution allows the network engineer to explicitly name each private network and assign the name to individual private WAN overlays to ensure tunnel initiation only happens within the common private network. As shown in Figure 4.5 a network engineer can navigate to `Private Network Names` and create new entries by going to `Configure > Network Services > Private Network Name > {New} > {Private Network Name} > {Save Changes}`

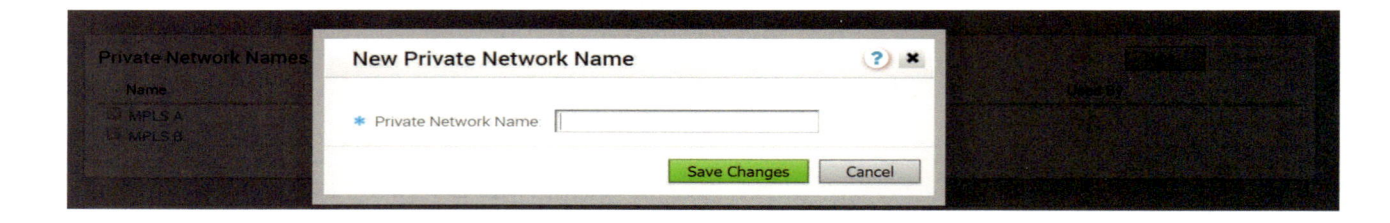

Figure 4.5: Create a new name for use with a specific private network

One Overlay per Physical Interface

Figure 4.6 shows a typical hybrid WAN branch. The VMware SD-WAN Edge terminates two WAN interfaces, one public and one private. Note the VMware SD-WAN Edge can terminate the MPLS link from the PE directly (copper or fiber Ethernet). In this scenario, there is a one-to-one relationship between the physical interface and WAN overlay. The physical interface itself can have 802.1Q VLAN tag as well.

Figure 4.6: One overlay per physical interface

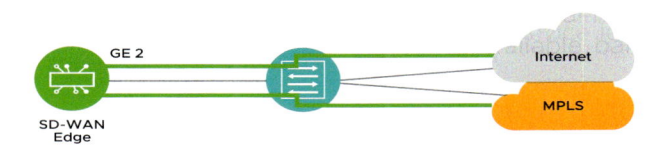

Multiple Overlays per Physical Interface

Figure 4.7 shows another scenario where the VMware SD-WAN Edge terminates only one physical interface. The upstream switchport is configured as trunk to pass multiple VLANs with different 802.1Q tags to the VMware SD-WAN Edge. Some service providers may choose to deliver both MPLS and internet over the same physical interface. The interface is configured with the IP address of the first VLAN (101), and the next-hop information to reach the internet. Next create a public user-defined WAN overlay with the same IP address, next-hop and VLAN ID. Then create a private user-defined WAN overlay and use the optional configuration in the WAN overlay for the private network IP address, next-hop and VLAN ID (102). Figure 4.8 shows a screenshot of the public user-defined WAN overlay. Figure 4.9 shows a screenshot of the private user-defined WAN.

Figure 4.7: Multiple overlays per physical interface

Topologies

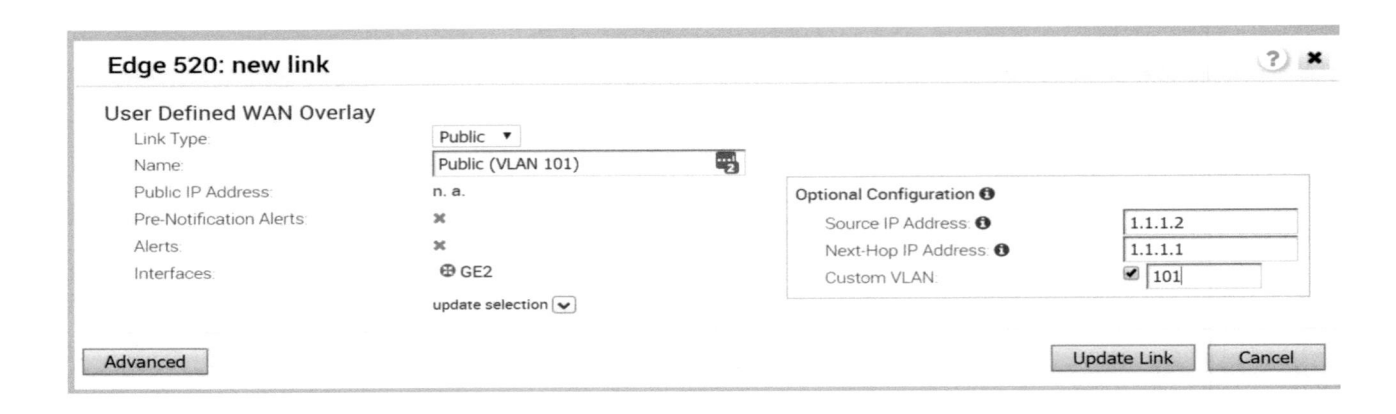

Figure 4.8: Public user-defined WAN Overlay

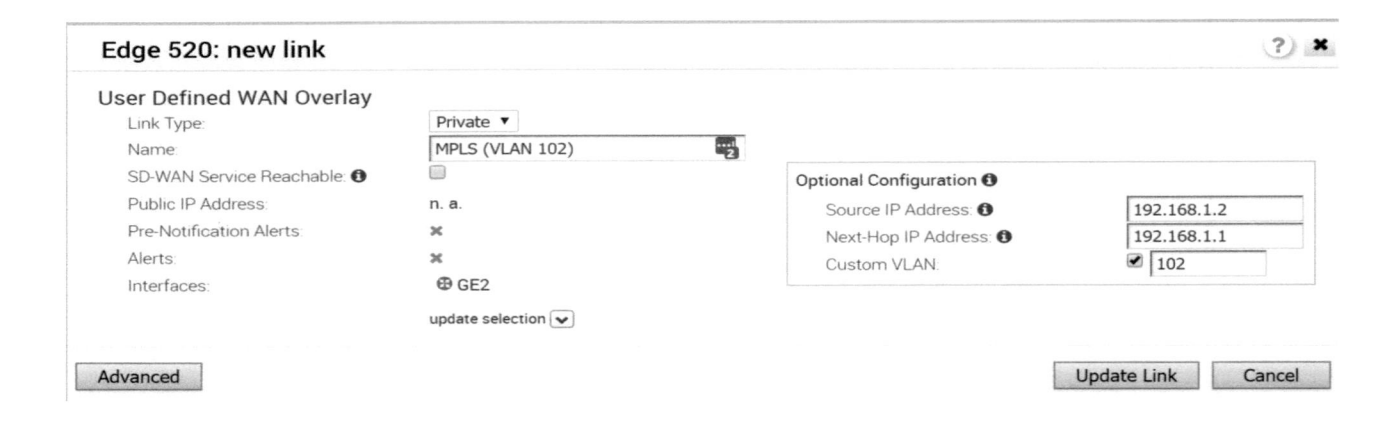

Figure 4.9: Private user-defined WAN Overlay

Key Takeaways:

- By default, all routed interfaces are configured to auto-detect the tunnel
- SD-WAN overlay on private links should be configured as user-defined
- Disable WAN overlay settings on LAN-facing routed interfaces
- VMware SD-WAN supports multiple SD-WAN overlay on a single routed interface
- Enable the **SD-WAN Service Reachable** feature for edges with private links for reachability to VMware SD-WAN Orchestrator and VMware SD-WAN Gateway on public internet
- To separate WAN overlays on a setup with private links from different ISPs, use the private network *name* option

Overlay Topologies

Overlay topologies are formed using the overlay types discussed in the previous chapter. This chapter will cover different topology types such as traditional hub-and-spoke or branch-to-branch.

The SD-WAN overlay can be built over any network that supports routing of IP packets, such as wired internet services, wireless internet services, private networks such as MPLS, along with point-to-point services such as those over radio, microwave, or fibre. SD-WAN overlays are built using VCMP tunnels.

By default, all VMware SD-WAN Edges in a VMware SD-WAN solution will build VCMP tunnels to a designated primary and secondary VMware SD-WAN Gateway. In addition, the SD-WAN overlay can be configured to build **Branch to Branch VCMP tunnels** and additional **VMware SD-WAN Edge to VMware SD-WAN Gateway VCMP tunnels** for access to **Non-VeloCloud Sites.** This is accomplished using IPSec tunnels from VMware SD-WAN Gateway to IPSec tunnel end-points. A **Non-VeloCloud Site** is any generic IPSec destination.

A Non-VeloCloud Site can be any generic IPSec destination, which should be made accessible from VMware SD-WAN Edges. This is accomplished using an IPSec tunnel from the VMware SD-WAN Gateway to the configured Non-VeloCloud Site.

The VCMP protocol works through NAT and implements a proprietary header for packet sequencing, time stamping, and detecting packet loss.

In order to create topologies, a network engineer will use the **Cloud VPN** feature. Figure 4.10 from the VMware SD-WAN Orchestrator User Interface provides an example of how to prepare VMware SD-WAN Edges in the same profile for SD-WAN overlay beyond the default VMware SD-WAN Edge to VMware SD-WAN Gateway tunnels, by using the **Cloud VPN slider**. This action, performed at the profile level in the VMware SD-WAN Orchestrator, instructs all VMware SD-WAN Edges in the same profile to enable control plane routing. To reach this section in the VMware SD-WAN Orchestrator UI, navigate to `Configure > Profiles > {Profile Name} > Device Tab > Cloud VPN`

Figure 4.10: VMware SD-WAN Orchestrator UI—Cloud VPN slider for enabling control plane routing

This base level cloud VPN configuration is typically applied at the profile for VMware SD-WAN Edges deployed into the data center (DC).

A **Hub** is an explicit role that's assigned to a VMware SD-WAN Edge. Hubs are typically located in the data center and they terminate multiple overlay tunnels. In the profile for a **Hub**, branch-to-branch VPN is typically not enabled as servers in each data center will leverage the data center interconnect (DCI) for communication rather than relying on VMware SD-WAN Edges in each DC to build tunnels to each other.

A separate profile is recommended for branch VMware SD-WAN Edges which will point to the **Hubs** via the `Branch to VeloCloud Hubs` configuration.

Hub-and-Spoke VPN

A VMware SD-WAN hub-and-spoke topology describes one or more branch VMware SD-WAN Edges aka **spokes**, builds one or more permanent tunnels to a centrally located edge in head office or the data center—known as the **Hub**. Figure 4.11 illustrates a typical hub-and-spoke topology with one VMware SD-WAN Edge configured as a Hub.

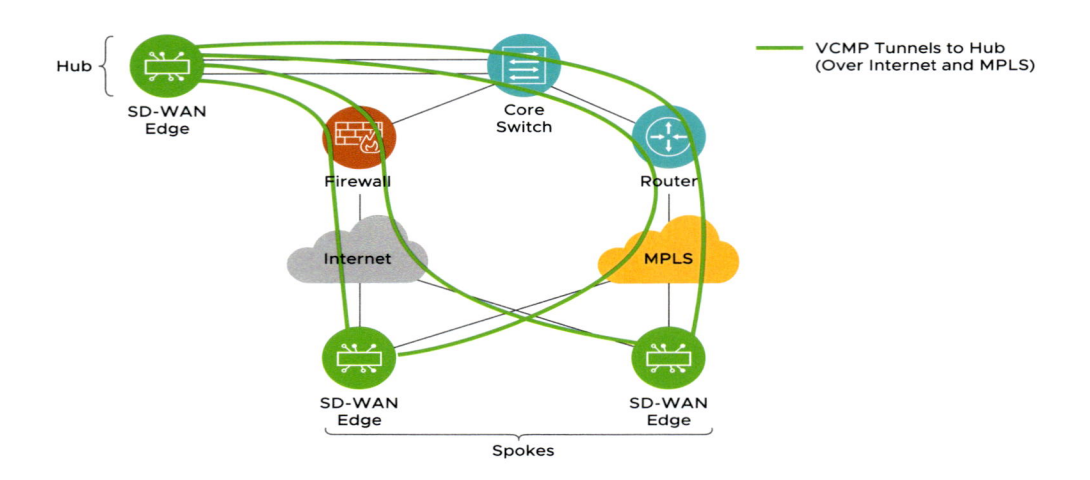

Figure 4.11: VMware SD-WAN—example of hub-spoke topology

Configuration for nominating a hub is typically done in the VMware SD-WAN Orchestrator using a profile with branch sites as members. In the same profile, `Branch to VeloCloud Hubs` is enabled and VMware SD-WAN Edges that are to be hubs for the branches in this profile are selected from a list. See Figure 4.12 for an example where two data center edges are being selected as hubs. Navigate to `Configure > Profiles > {Profile Name} > Device Tab > Cloud VPN > Branch to VeloCloud Hubs > Enable`

This configuration will instruct all VMware SD-WAN Edges that are a member of this profile to build permanent tunnels to `DataCenter1` and `DataCenter2` VMware SD-WAN Edges.

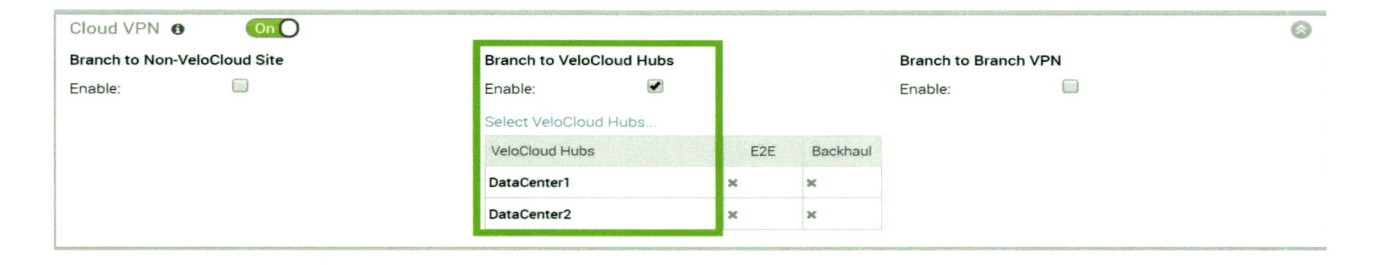

Figure 4.12: VMware SD-WAN Orchestrator profile with hub selection

Branch to Branch VPN

VMware SD-WAN can be configured to allow branch-to-branch communication. The traffic path for branch-to-branch can either be via:

- VMware SD-WAN Gateways (selected by default); or
- VMware SD-WAN Edges configured as hubs - selected in the profile under `Branch to Branch VPN`
- VMware SD-WAN Edges can also be configured to utilize `Dynamic Branch to Branch VPN` as covered in Figure 4.13

Note that VMware SD-WAN Gateways can be used to stitch tunnels together, i.e. route between them, in the same fashion as a hub. In hybrid deployments where internet and MPLS underlays are in use, a hub is a better choice for routing between tunnels as a hub will typically have connectivity to both underlay types.

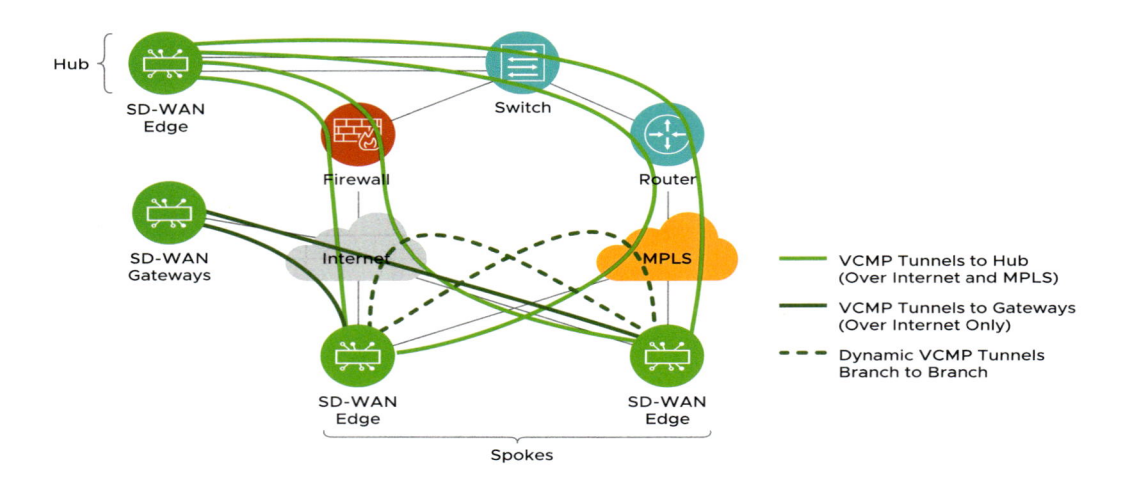

Figure 4.13: Example of Branch to Branch VPN via either a hub, gateway, or Dynamic Branch to Branch VCMP Tunnel

When you enable **Dynamic Branch to Branch VPN**, the first few packets will go through the cloud-hosted VMware SD-WAN Gateway or the Hub. If the initiating VMware SD-WAN Edge determines that traffic can be routed branch-to-branch direct using a VCMP tunnel between VMware SD-WAN Edges, then a direct dynamic tunnel is created between the branches. Once the tunnel is established, traffic begins to flow over the VCMP tunnel between the branches. In the event dynamic branch to branch tunnels fail to establish (e.g. if one of the edges is behind symmetric NAT), traffic will continue to flow via VMware SD-WAN Gateway or the hub.

Figure 4.14 shows the traffic path transiting the Hub (dark blue) or transiting the cloud-hosted VMware SD-WAN Gateway (light blue), then the branch-to-branch direct path if enabled (blue dotted line). After an idle time out (no traffic between branches) of 3 minutes, the initiating VMware SD-WAN Edge will trigger a teardown of the dynamic tunnel.

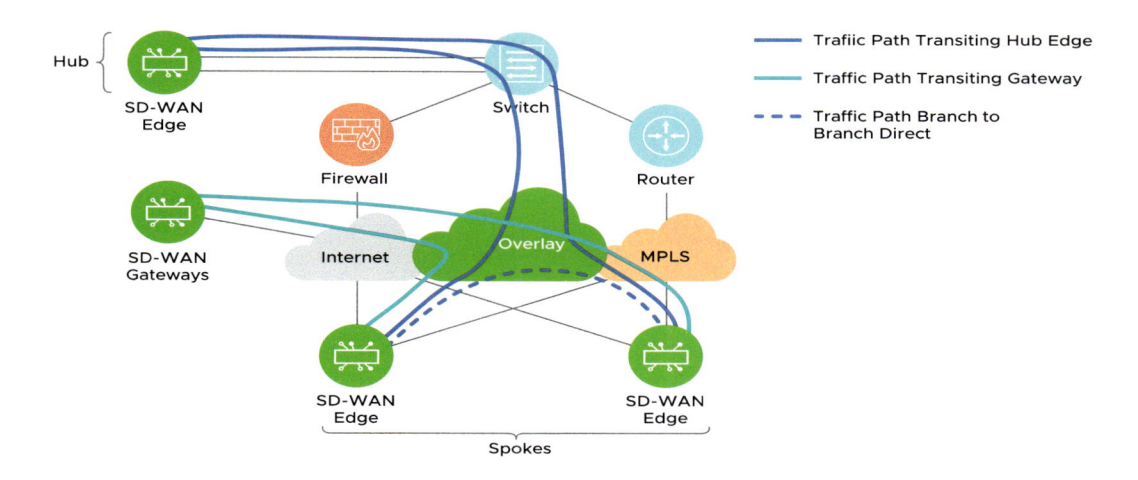

Figure 4.14: Traffic flow for branch-to-branch via either hub edge, gateway, or direct

To configure branch-to-branch, the `Cloud VPN` service is turned on in the `profile` which has branch edges as members, `Branch to VeloCloud Hubs` are selected, and the `Branch to Branch VPN` is then checked. As seen in Figure 4.15, there is an option to select `Cloud Gateways or VeloCloud Hubs` as the transit device for branch-to-branch communication. Navigate to `Configure > Profiles > {Profile Name} > Device Tab > Cloud VPN > Branch to Branch VPN > Enable` and `> Dynamic Branch To Branch VPN > Enable`

Cloud VPN 🛈	**On** ⬤		

Branch to Non-VeloCloud Site
Enable: ☐

Branch to VeloCloud Hubs
Enable: ☑
Select VeloCloud Hubs...

VeloCloud Hubs	E2E	Backhaul
DataCenter1	☑	✕
DataCenter2	☑	✕

Branch to Branch VPN
Enable: ☑
Isolate Profile 🛈 ☐

Use Cloud Gateways: ○
Use VeloCloud Hubs for VPN ⬤
Select VeloCloud Hubs... 🛈

Branch to Branch VPN Hubs	order
DataCenter1	1
DataCenter2	2

Dynamic Branch To Branch VPN
Enabled ☑
  To All Edges ⬤
  To Edges Within Profile ○

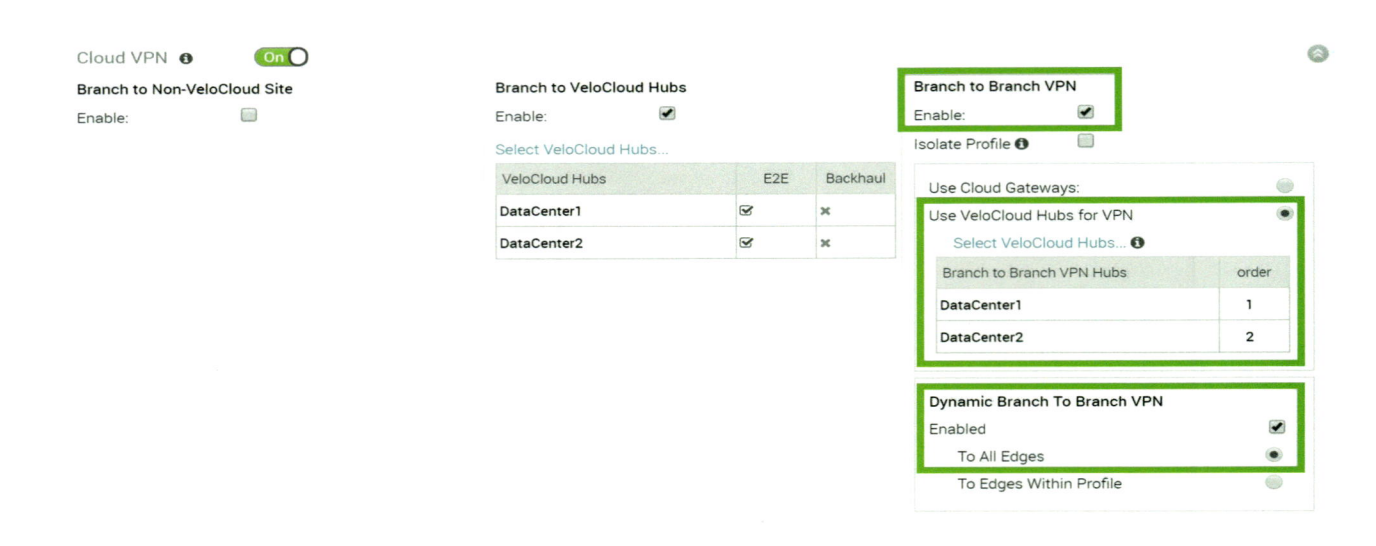

Figure 4.15: VMware SD-WAN Orchestrator profile with Dynamic Branch To Branch VPN enabled

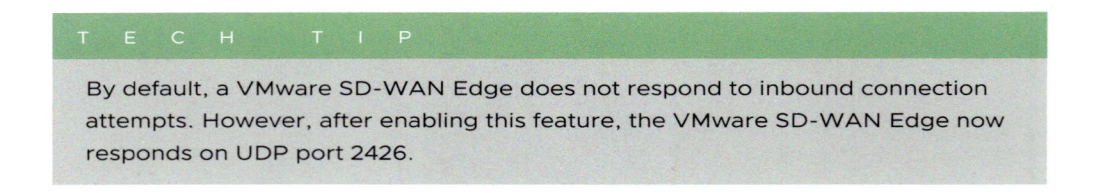

> **T E C H   T I P**
>
> By default, a VMware SD-WAN Edge does not respond to inbound connection attempts. However, after enabling this feature, the VMware SD-WAN Edge now responds on UDP port 2426.

Branch to Branch VPN using Profile Isolation

VMware SD-WAN can be configured to allow branch-to-branch communication between branches from different profiles as long as they are connected to a common VMware SD-WAN Gateway or Hub. If a customer wants to prevent a branch VMware SD-WAN Edge associated with a given profile from connecting to another branch associated with a different profile, VMware SD-WAN offers **Segmentation** and **VPN Profile Isolation**. The recommended method is to use the SD-WAN segmentation feature to assign VMware SD-WAN Edges to different segments. Refer to Chapter 6: *Security*, section *VMware SD-WAN Segmentation*, for more information on this feature.

For customers who prefer to not enable the segmentation feature or who want to achieve VMware SD-WAN Edge isolation within one segment, the `Isolate`

`Profile` can be used (see Figure 4.15 again).

When **VPN Profile Isolation** is enabled for a profile, the VMware SD-WAN Edges within that profile will only learn:

- Routes to VMware SD-WAN Edges within its own profile.
- Routes to the assigned Hubs as well as underlay routes learned by the Hub.

Cloud VPN for Non-VeloCloud Site

There are several situations where a site is not yet able to migrate to SD-WAN. Examples would be during migration, or during a merger and acquisition, where not every site can be moved to SD-WAN in a single day. The other example would be a third-party provider that doesn't have VMware SD-WAN Edges installed locally. The site that does not run SD-WAN is called a **Non-VeloCloud Site (NVS)**. Figure 4.16 shows a typical topology of connecting to a NVS. The VMware SD-WAN Gateway forms a standard IPSec tunnel to the device in the NVS, represented in Figure 4.16 by a traditional data center. From the branch VMware SD-WAN Edge perspective, it forms a VCMP tunnel to the VMware SD-WAN Gateway and learns the prefixes belonging to the traditional data center. As a result, the VMware SD-WAN Edge is able to connect to the traditional data center.

Figure 4.16: Example of VMware SD-WAN Edge to a Non-VeloCloud Site (NSV). Connection is via the VMware SD-WAN Gateway.

Compared with having the VMware SD-WAN Edge directly form a standard IPSec tunnel to the traditional data center, the approach of utilizing the VMware SD-WAN Gateway is much more scalable as it dramatically reduces the number of tunnels needed from individual branches. In order to reach the screen seen in Figure 4.17, navigate to `Configure > Profiles > {Profile Name} > Device Tab > Cloud VPN > Branch to Non-VeloCloud Site > Enable` A configuration example for NVS can be seen in Figure 4.17.

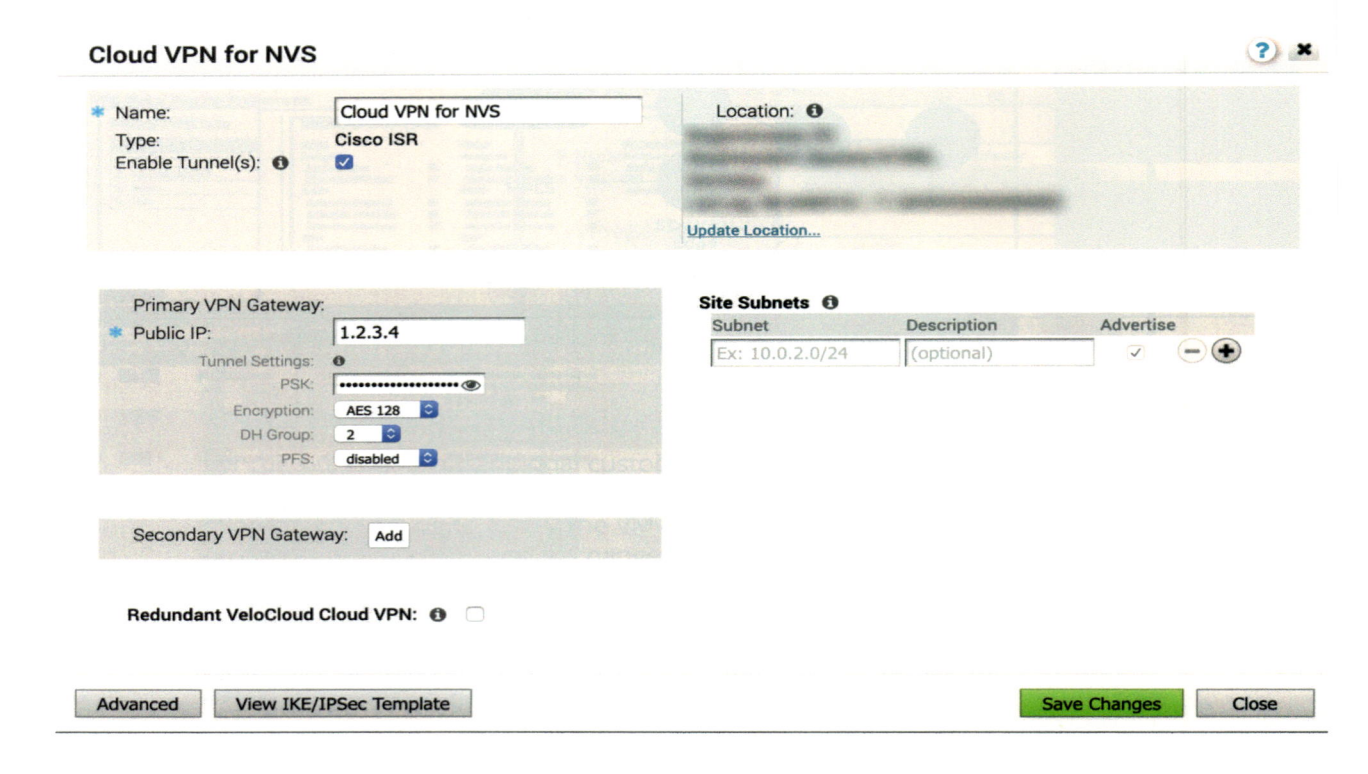

Figure 4.17: Configure Cloud VPN for NVS

> **Key Takeaways:**
>
> - By default, all VMware SD-WAN Edges in a VMware SD-WAN solution will build VCMP tunnels to a designated primary and secondary VMware SD-WAN Gateway.
> - In addition, the SD-WAN overlay can be configured to build branch-to-branch VCMP tunnels and additional VMware SD-WAN Edge to VMware SD-WAN Gateway VCMP tunnels for access to Non-VeloCloud Sites.

Deployment Models

Overview

The VMware SD-WAN solution enables organizations to connect to cloud-hosted applications using an agile, transport-independent overlay that ensures private network performance, reliability, and manageability. There are different approaches that can be taken for the hosting of the solution itself, which are detailed in the following.

Enterprise Deployment Model (VMware Cloud Hosted)

VMware Enterprise Topology (aka OTT: Over The Top) makes use of VMware SD-WAN Gateways and VMware SD-WAN Orchestrators co-located at major facilities around the world to help enable the cloud-delivered strategy. This strategy leverages proximity of the VMware SD-WAN Gateways to XaaS hosting locations to remediate last-mile connectivity issues commonly seen at the branch. This enables support of the shift to cloud-based workloads and applications in place of doing traditional backhaul to the data center.

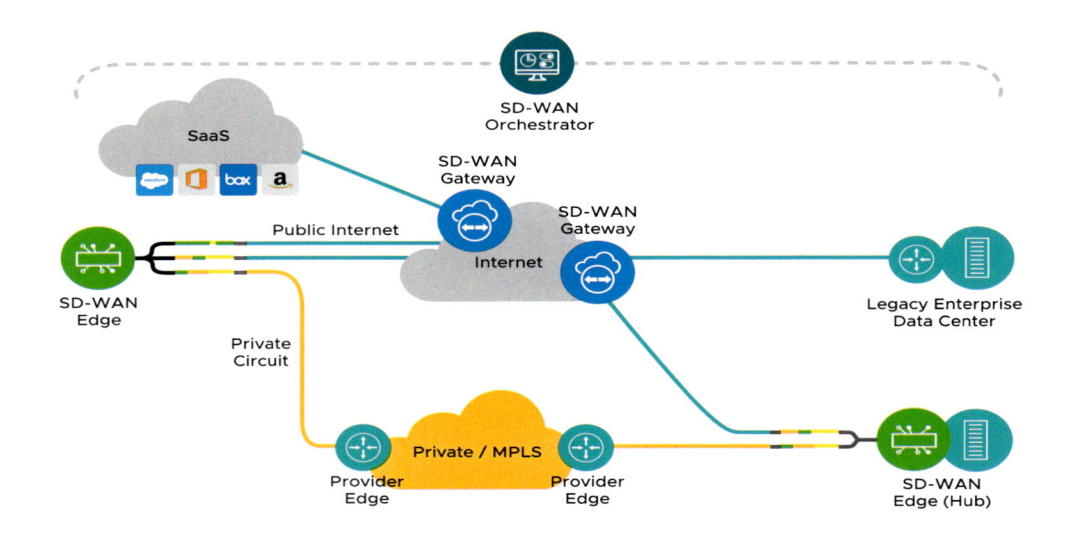

Figure 4.18: VMware SD-WAN Over The Top architecture with cloud-hosted gateway and orchestrator

The VMware SD-WAN Orchestrators are also hosted in cloud locations across the world, providing ready access. To further insulate from extended or significant outages in any one region, each VMware SD-WAN Orchestrator is protected with a disaster recovery peer hosted in a different geographic region. As the VMware SD-WAN Orchestrator only participates in the management plane, any outage experienced here would have no control or data plane impact on VMware SD-WAN Gateways or VMware SD-WAN Edges connected to it. Visibility and manageability of these devices could be impacted, however, and so a redundancy strategy is in place to mitigate any potential impact.

The VMware SD-WAN Gateways in the cloud-hosted model are assigned automatically via geolocation, with a primary and geo-redundant secondary gateway allocated per VMware SD-WAN Edges. In organizations where cloud VPN is enabled, a super gateway and redundant super gateway are also elected to ensure that control plane resiliency is maintained. A super gateway is used as gateway of last resort for VPN exchange point between branch sites in different regions as depicted in Figure 4.19. This super gateway is assigned by the VMware SD-WAN Orchestrator at the approximate geographic center of the organization.

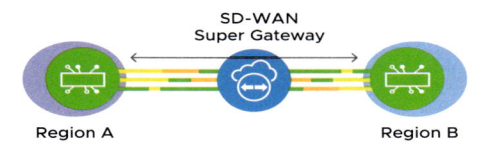

Figure 4.19: VMware SD-WAN super gateway

Service Provider Deployment Model (Provider-hosted or On-premises Deployment)

The service provider-hosted topology is very similar to the VMware cloud-hosted topology as it consists of the same basic components. In this model the VMware SD-WAN Orchestrator and VMware SD-WAN Gateway will be located at service provider-owned and operated facilities. Additional aspects of the service provider deployment model such as partner gateways, are covered in a later section, Chapter 8: *Service Provider*.

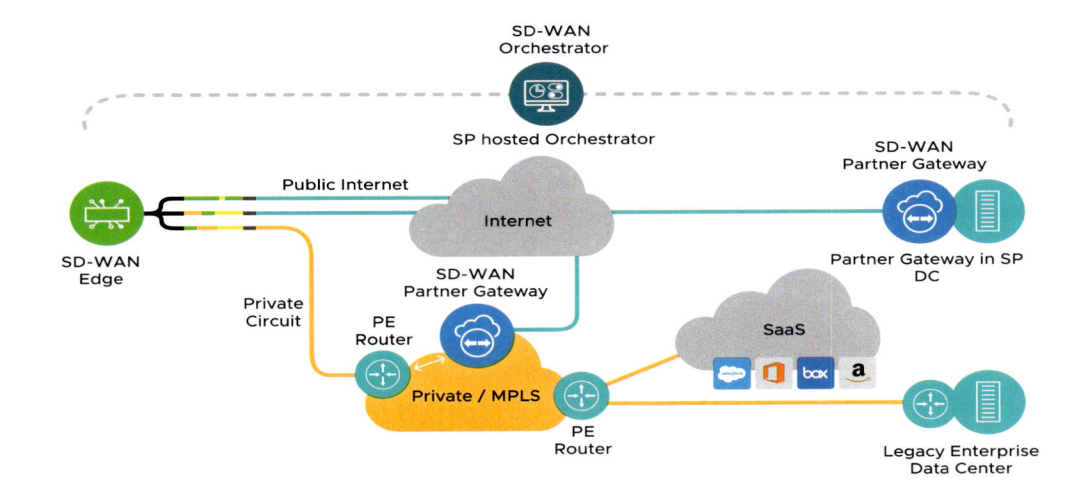

Figure 4.20: VMware SD-WAN architecture for the service provider-hosted offering

Topologies   71

One of the main advantages in the service provider-hosted model is in the function of the gateway itself. This partner gateway functionality provides an advantage to the service provider in that they are hosting the gateway themselves, providing last-mile remediation benefits to their various service offerings.

On-premises

The on-premises hosted topology differs in that the VMware SD-WAN Gateway and VMware SD-WAN Orchestrator are no longer cloud-hosted components. In these deployments, the VMware SD-WAN Gateway will either act as a controller or will carry both control and data plane traffic. In on-premises scenarios, where partner gateway mode is utilized, the configuration and topologies, even use cases, would closely mirror those found in a service provider-hosted topology.

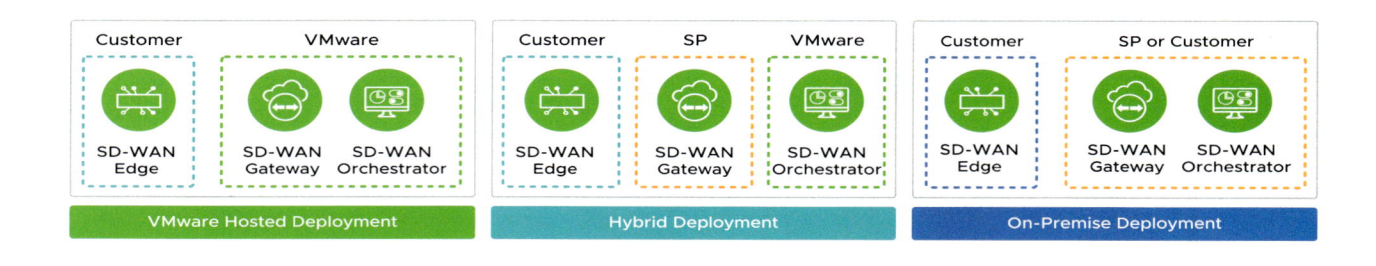

Figure 4.21: VMware SD-WAN deployment options

Key Takeaways:

- VMware SD-WAN has a nice simplified deployment model for organizations whereby they only have to install the VMware SD-WAN Edge and get the control plane and management plane from the VMware hosted Orchestrator and Gateway.
- Service providers can deploy their own VMware SD-WAN Orchestrator and VMware SD-WAN Gateway for multiple customers and optimize the customer traffic. This gives SD-WAN benefits for the last-mile and access to private networks for the mid-mile.
- Organization-hosted or on-premises deployment requires that the organization or service provider install the VMware SD-WAN Edges, Orchestrator and Gateways in their own branches and data centers and operate management plane and control plane directly.

Site Topology and Redundancy

This chapter covers the commonly deployed topologies for VMware SD-WAN. It describes the network topology, deployment considerations, and best practices when inserting VMware SD-WAN Edge into a site. Figure 4.22 is a high-level diagram of the network topology showing the individual site topologies covered in this chapter. It shows a single data center and 5 branch sites. There are two underlay transports shown—one MPLS and one internet service provider. The VMware SD-WAN Orchestrators and Gateways are VMware cloud-hosted and reachable by internet transport. The VMware SD-WAN Edges will attempt to reach the VMware SD-WAN Orchestrators and Gateways and establish control plane connections over each transport.

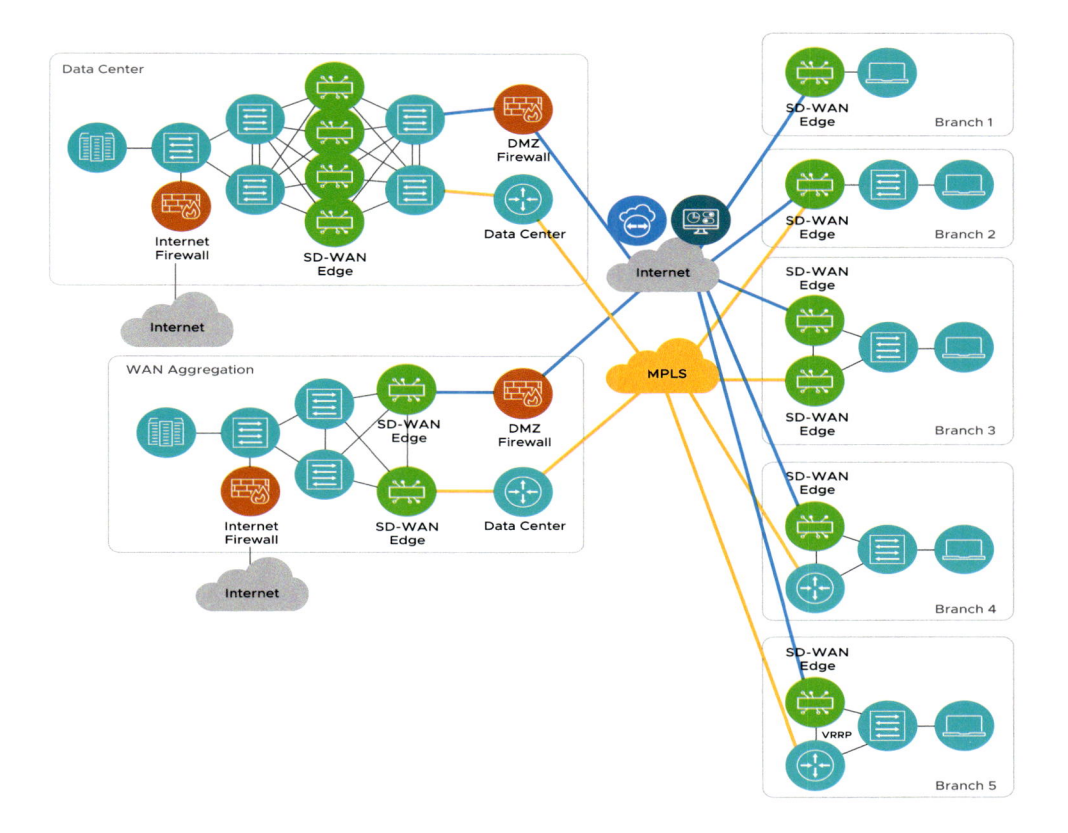

Figure 4.22: Topology and High Availability overview

VMware SD-WAN Edge Redundancy Options

There are three different design options for redundancy designs:

- High Availability (used in branch and/or data center)
- Clustering (used in data center only)
- VRRP (used in branch only)

Figure 4.23 shows the `High Availability` options in the UI, which can be reached by `Configure > Edges > {Edge} > Device tab > High`

Topologies 75

`Availability`. The first two options will be discussed in the data center context.

High Availability

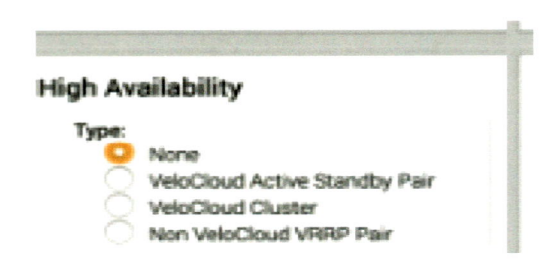

Type:
- None
- VeloCloud Active Standby Pair
- VeloCloud Cluster
- Non VeloCloud VRRP Pair

Figure 4.23 High Availability options

Data Center Design Options

For data center locations, there are two designs that are commonly used. They are:

- High Availability (HA) mode
- Clustering mode

There are distinct characteristics that are associated with each option. A summary comparison between the two modes is shown in Table 4.1.

High Availability Mode	Clustering Mode
High availability for sites with a single pair of VMware SD-WAN Edges. Branches or small scale data centers.	Recommended for data center designs, typically not applicable to branch designs.
Limited to one pair of VMware SD-WAN Edges.	Allow horizontal scaling of VMware SD-WAN Edges to meet performance requirements. Adding VMware SD-WAN Edges in a cluster is simple and not disruptive.
LAN side connection to downstream devices can leverage static routes or dynamic routing protocols (eBGP).	LAN side connection to downstream devices requires dynamic routing protocol, like eBGP
Capacity is limited to the active device.	Capacity is aggregated across all the VMware SD-WAN Edges in the cluster.

Table 4.1: High Availability / Cluster comparison

Each design scenario is discussed in more detail in the following section.

High Availability Mode

VMware SD-WAN Edges configured with HA mode end up with exactly the same configuration. On the VMware SD-WAN Orchestrator, they will show up as a single VMware SD-WAN Edge device. There are two options for configuring HA mode, and the VMware SD-WAN Edge will automatically select either option depending on how the WAN interfaces are connected. Forming an HA pair requires both devices to be of the same type.

Option 1 - Standard HA Mode

Figure 4.24 provides an overview of Standard HA Mode. In this design, the VMware SD-WAN Edges connect to different switches on the WAN and LAN interfaces. The two VMware SD-WAN Edges must have physical mirror connections on both the WAN and LAN interfaces. The VMware SD-WAN Edges, one active and one standby, are physically connected back-to-back to establish a failover link. The standby VMware SD-WAN Edge blocks all ports except the HA port for the failover link. HA link is used for sending keepalive and communication of active / standby state between the two edges.

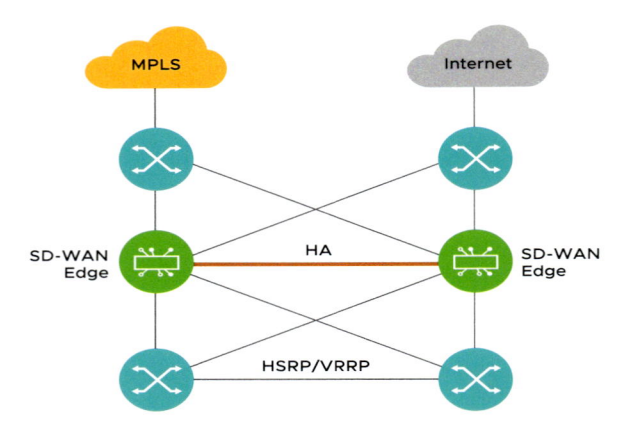

Figure 4.24: Standard HA mode

Deploying High Availability (HA) allows for hitless upgrades.

Full redundancy is achieved by connecting the VMware SD-WAN Edges to redundant LAN switches. Typically, they are L3 switches in HSRP/VRRP configuration. In order to achieve end-system connectivity on the LAN side, the VMware SD-WAN Edge pair needs to have a static route pointing towards the VRPP/HSRP Virtual IP address configured on the switches, or a dynamic routing protocol (either BGP or OSPF).

On the WAN side, there must be either a WAN modem with two ports to connect both VMware SD-WAN Edges in the HA pair, or there must be an additional pair of switches providing the WAN connectivity to both VMware SD-WAN Edges in the HA pair. For example, Figure 4.24 shows a hybrid site with an MPLS and an internet connection, which are provided to both VMware SD-WAN Edges through these switches.

The failover from active to standby happens when the edges detect that physical connectivity has been lost on either the WAN or the LAN connections.

Option 2 - Enhanced HA Mode

Figure 4.25 shows Enhanced HA Mode. It eliminates the need for layer 2 switches on the WAN side. To enable this option, the WAN connections **must** terminate on different interface port numbers (see `GE2` and `GE3` on figure 4.25). When the active VMware SD-WAN Edge detects different WAN link(s) connected to the standby edge when compared to the link(s) connected to itself, it will automatically select enhanced HA mode.

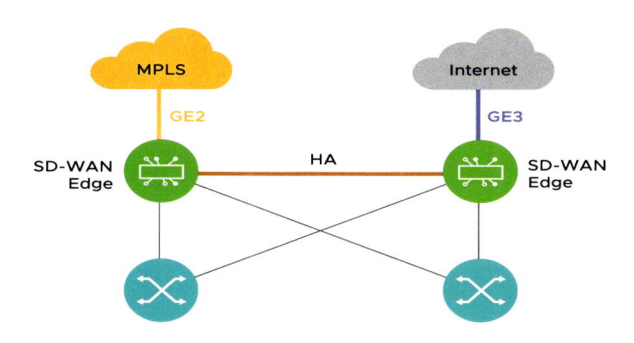

Figure 4.25: Enhanced HA Mode

For the active VMware SD-WAN Edge to leverage the WAN link connecting to the standby VMware SD-WAN Edge, the active VMware SD-WAN Edge establishes an **SD-WAN Overlay** through the HA link. Traffic from the LAN side is forwarded to the active VMware SD-WAN Edge, and the business policy defined will determine how the traffic flows are distributed across the links on both VMware SD-WAN Edges.

In the event of a failure of the heartbeat link between the active and standby VMware SD-WAN Edges, split brain protection would occur automatically, as the VMware SD-WAN Gateway acts as a witness. In this scenario, the VMware SD-WAN Edge which used to be standby, and is now active, would build a tunnel to the same VMware SD-WAN Gateway to which the currently active VMware SD-WAN Edge in the pair has already established tunnels. When the VMware SD-WAN Gateway compares the serial number of both VMware SD-WAN edges, it will notice the 'new' active VMware SD-WAN Edge, and it will instruct the VMware SD-WAN Edge which used to be active before the failover event to go into standby mode.

> All High Availability modes are applicable to branch site deployments.

Cluster Mode

Cluster mode allows the deployed VMware SD-WAN Edges in the data center to be used in a resource-aware cluster solution. It completes the common redundancy in the data center switch fabric with an equally redundant SD-WAN cluster solution. Each new connection from a VMware SD-WAN Edge is dynamically assigned by the VMware SD-WAN Gateway to the least busy VMware SD-WAN Edge in the cluster. An overview topology is shown in Figure 4.26.

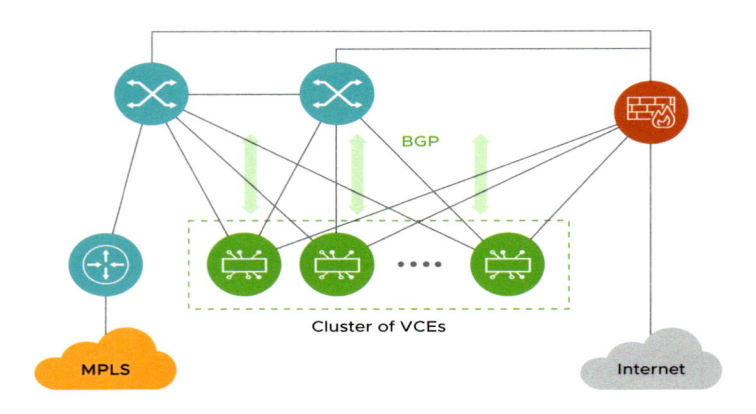

Figure 4.26: Hub Cluster design

In this mode, the VMware SD-WAN Edges are configured to join a hub cluster configuration. All edges in the hub cluster are presented as one hub to the branch edges. The VMware SD-WAN Gateway determines the edge in the hub cluster to which a branch edge will build an overlay tunnel. The VMware SD-WAN Gateway selects one of the edges in the hub cluster, based on lowest utilization, and assigns it to a branch edge. This is shown in Figure 4.27.

Every hub in the hub cluster reports usage and load statistics to the VMware SD-WAN Gateway periodically. The VMware SD-WAN Gateway maintains a list of hubs sorted by load. The branch VMware SD-WAN Edge requests the VMware SD-WAN Gateway for the hub IP address. The VMware SD-WAN Gateway assigns the least loaded hub to the branch VMware SD-WAN Edge. No state is synced between the hubs in the cluster. The tunnel is set up by the VMware SD-WAN Edge to the assigned Hub.

Here are the steps articulated in Figure 4.27.

1. Each hub reports usage and loads stats to the NSX SD-WAN Controller (VCC) periodically. VCC maintains a list of hubs in an increasing order of their load.
2. Branch VCE requests VCC for hub IP address
3. VCC assigns leasts loaded hub to the branch VCE
4. There will be no state sync between the hubs in the cluster
5. Branch VCE sets up tunnel to the assigned hub.

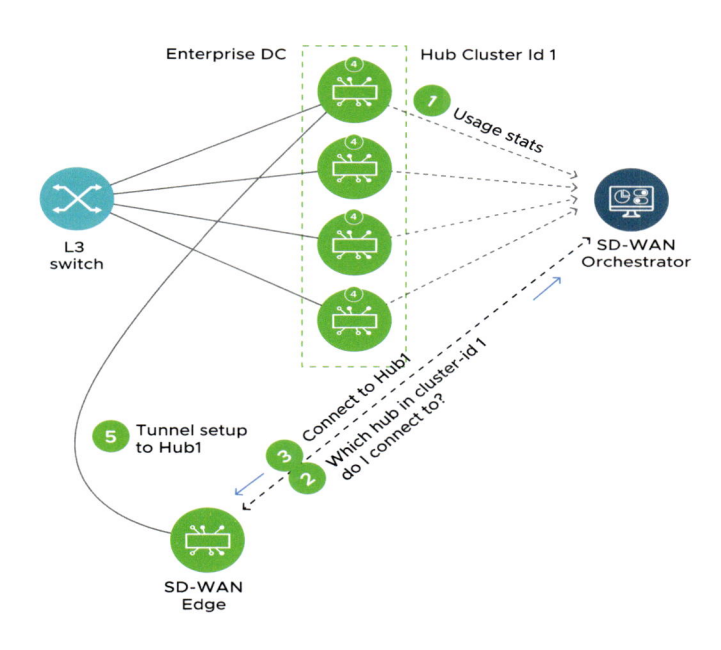

Figure 4.27: Resource-aware hub clustering

There is a configuration option in the VMware SD-WAN Orchestrator that would allow the VMware SD-WAN Gateway to automatically rebalance the connections over time. The VMware SD-WAN Edges in the cluster do not communicate with other VMware SD-WAN Edges in the same cluster.

Each cluster member will have its own IP addressing for the WAN and LAN interfaces. All the VMware SD-WAN Edges in the hub cluster are required to run a dynamic routing protocol, e.g. eBGP, with the layer 3 device(s) on the LAN side with a unique Autonomous System Number (ASN) for each cluster member. This allows the VMware SD-WAN Edges in the hub cluster to advertise the LAN prefixes into the overlay tunnels and draw traffic into the data center. When a VMware SD-WAN Edge in the hub cluster loses its routing adjacency on the LAN side, the VMware SD-WAN Orchestrator removes the VMware SD-WAN Edge in question from the list of available hubs and re-balances the VCMP tunnels on that device to the remaining cluster members.

Hub clustering is used for horizontal scale and redundancy for VMware SD-WAN Edges in the same data center location. Hub clustering is recommended over HA for data center deployment in most cases. At least N+1 hubs are recommended in a cluster, assuming N is the minimum number of hubs needed to meet the bandwidth and tunnel requirements.

Branch Details

There are several branch topologies commonly deployed using VMware SD-WAN. They are:

- Single VMware SD-WAN Edge with internet-only connection.
- Single VMware SD-WAN Edge with MPLS and internet connections.
- Dual VMware SD-WAN Edges with MPLS and internet connections.
- Single VMware SD-WAN Edge and a traditional router in-path.
- Single VMware SD-WAN Edge and a traditional router in a VRRP configuration.

The following will cover the VMware SD-WAN insertion topologies.

Single VMware SD-WAN Edge with internet-only connection

The internet-only sites will build a SD-WAN Overlay to the hubs using internet transport. If dynamic branch-to-branch VPN is enabled in the profile for the sites, the sites will build on-demand overlay tunnels to other sites connecting to the internet transport. To reach sites that are only connected to MPLS transport, the traffic from an internet-only site will have to go through the data center or hub location. Figure 4.28 shows this topology.

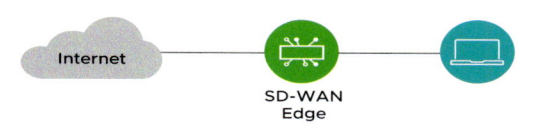

Figure 4.28: Edge with internet-only connection

Single VMware SD-WAN Edge with MPLS and internet connections

Sites using a single VMware SD-WAN Edge with both MPLS and internet connections will build a SD-WAN Overlay to the hubs using both MPLS and internet transports. When the Hub is used as the transit location for branch-to-branch communication, the VMware SD-WAN Edge at the branch locations is not required to run dynamic routing protocols such as BGP with the provider edge router on the MPLS interface. The only requirement for the VMware SD-WAN Edge at the branch location is to have a default route on the WAN interface that points the traffic to the provider edge router. Figure 4.29 shows this topology.

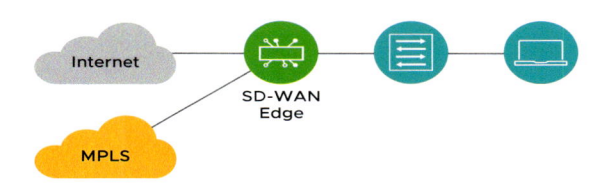

Figure 4.29: Edge with MPLS and internet connection

Dual VMware SD-WAN Edges with MPLS and internet connections

Sites using a dual VMware SD-WAN Edge pair with MPLS and internet connection will have the same design possibilities as the high-available examples discussed at the beginning of this chapter under *Data Center Design Options* (standard HA mode, enhanced HA mode). Figure 4.30 shows this topology.

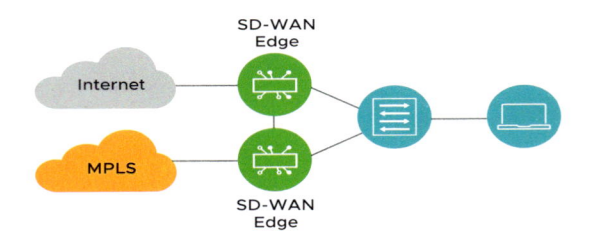

Figure 4.30: Dual Edge with MPLS and internet connection

Single VMware SD-WAN Edge and a traditional router Off-Path

For the sites using a traditional customer equipment (CE) router to terminate WAN connections, one VMware SD-WAN Edge can be inserted and take advantage of dual circuits coming to a site. The VMware SD-WAN Edge will run a dynamic routing protocol such as BGP, and exchange routing information with the CE router. The CE router will advertise the VMware SD-WAN Edge interface address into the MPLS network, thus providing reachability to the edge for other parts of the MPLS network. The VMware SD-WAN Edge can then form the SD-WAN Overlays through both transports concurrently. Figure 4.31 shows this topology.

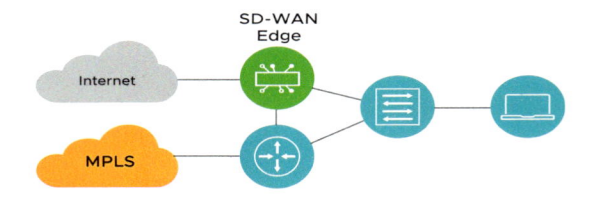

Figure 4.31: Edge and traditional router in-path

Single VMware SD-WAN Edge and a traditional router in a VRRP configuration

This topology is the same as the previous one, but now VRRP is active between the traditional router and the VMware SD-WAN Edge. The VMware SD-WAN Edge is the VRRP master and provides benefits from the SD-WAN Overlay functionalities in normal conditions. In case of a VMware SD-WAN Edge failure, the traditional router is available as a fallback option.

SD-WAN 1:1

This setup is common for migration scenarios and not commonly used in day-to-day operations.

> **Key Takeaways:**
>
> - At the data center, the customer has the choice with VMware SD-WAN to choose between **High Availability Mode** with up to two edges or **Cluster Mode** with up to eight edges in a cluster.
> - VMware SD-WAN brings a lot of flexibility to the branch. All variants from one edge with one or two links up to two edges with internet and MPLS links in a redundant mode are possible. Migration scenarios are fully supported.

After a deep-dive design session with Raycliff the VMware SD-WAN specialist SE, Rodney has begun to really wrap his mind around how to get started. Even if Allofu hits Alvina's very aggressive 3-year growth plan, Rodney is confident that the VMware SD-WAN can readily meet their network growth needs. Rodney is very intrigued by what he has learned and it's a bit of a mindset shift to manage his SD-WAN network via a hosted model, much less a GUI! His past experiences managing his network through fancy GUIs has been less than stellar. Raycliff shared that VMware has numerous customers that are adding broadband and even LTE connections to augment their private links. This gives their customers even more available bandwidth and is quite affordable. Rodney remains pessimistic about sending any business-critical traffic across the internet. He understands that it's encrypted, so that isn't his main concern, though he knows Hans the security architect will certainly have something to say. How can real-time traffic work reliably over the internet? Rodney thinks back to how well he and his buddies were doing on Fortnite last night and then Boom—Lag! Ugh, that was it and his wife just didn't understand his frustration. He wasn't really yelling, just excited and it's not like she hasn't seen every one of those silly cooking shows before anyway. He needs to dig into how VMware protects business traffic over lossy links and if this works as advertised, he might need an VMware SD-WAN Edge for home!

Alvina walks by the conference room and sees Rodney and the VMware SD-WAN specialist SE scribbling away on the whiteboard. What kind of name is Raycliff anyway? Alvina doesn't really care as long as they get this project moving forward. Rodney told Alvina that the plan is to kick off the pilot in North Dakota on first of the month, and that's only a week away.

5. Application Performance, Routing, and Cloud Access

Rodney has a meeting with Raycliff to understand how VMware's SD-WAN provides protection for applications. Rodney thinks to himself that needing to learn yet another Quality of Service (QoS) process is about as gratifying as hearing that he should stop drinking three extra-large double-shot latte's every morning! Rodney simply doesn't need such negativity in his life. Now that Rodney is sufficiently caffeinated, it's time to dig into these concepts. These types of tools always seem to be overly complex and frankly Rodney questions if anyone truly understands these implementations in his traditional network as it's different on every box! Alvina is hovering nearby, clearly anxious for Rodney to move faster. Rodney is certain that Alvina is making up excuses to keep passing by the conference room to monitor their progress.

Application Performance—Dynamic Multi-Path Optimization

With the need for IT to become more agile and more application-centric, the traditional mechanisms to ensure reliable forwarding of data need to be rethought. Traditional WAN QoS relies on Differentiated Service Code Point (DSCP) marking. Knowing that the internet does not carry DSCP markings, this approach has a few shortcomings as applications are moving towards the cloud and companies are migrating to SD-WAN with hybrid transports.

SD-WAN introduces a new paradigm in which the QoS mechanisms will need to be application and link quality aware. VMware SD-WAN **Dynamic Multi-Path Optimization (DMPO)** technology delivers assured application performance and uniform QoS mechanism across different transports.

The following provides more information on DMPO technology. DMPO provides optimizations between two VMware SD-WAN Edges or between a VMware SD-WAN Edge and a VMware SD-WAN Gateway.

There are 4 key functionalities of DMPO—continuous monitoring, dynamic application steering, on-demand remediation, and application-aware overlay QoS.

Continuous Monitoring

- **Automated bandwidth discovery**
  When a new internet WAN link is detected, the VMware SD-WAN Edge builds VCMP tunnels automatically to one or more VMware SD-WAN Gateways and performs bidirectional bandwidth tests against the closest VMware SD-WAN Gateway to determine the available bandwidth of the link. A similar process takes place between VMware SD-WAN Edge in a branch and VMware SD-WAN Edges configured as Hub. The bidirectional bandwidth for each path, as defined by a link-to-link pairing between two VMware SD-WAN Edges, is then automatically calculated. The VMware SD-WAN QoS scheduler shapes the overall traffic according to the available bandwidth of each path, which prevents congestion of the link.

- **Continuous path monitoring**
  Using the sequence numbers and time stamps in the VCMP header, DMPO performs unidirectional path monitoring in real-time, which provides continuous link quality metrics (aka latency, jitter, and packet loss) for all paths between any two VMware SD-WAN Edges or between a VMware SD-WAN Edge and a VMware SD-WAN Gateway. This allows independent steering decisions for upstream and downstream traffic, based on the link quality in each direction. DMPO performs both passive (using VCMP header information of user traffic) and/or active (VCMP probes in the absence of user traffic) monitoring.

Dynamic Application Steering

- **Application-Aware Per-Packet Steering**
  The Deep Application Recognition (DAR) engine in VMware SD-WAN Edges supports classification of traffic based on Layer 2 to Layer 7 attributes, which recognizes 3,000+ applications. DMPO provides per-packet steering decisions based on application characteristic and link quality, which allows a single traffic flow to be steered to another link mid-flow with no impact to user experience. Leveraging the continuous path monitoring capability, DMPO provides sub-second link failure and link degradation protection for all user traffic.

- **Bandwidth aggregation**
  DMPO allows a single traffic flow to fully utilize the available bandwidth by intelligently aggregating bandwidth across multiple links. Taking into account the applications' latency requirements and real-time link latency and available bandwidth, DMPO aggregates all available bandwidth for a single flow without delivering out-of-order.

On-Demand Remediation

Continuous path monitoring and application-aware per-packet steering combined provide sub-second link failure and link degradation protection when at least one good quality link is available. In the event that all links are suffering quality issues (e.g. jitter or packet loss), or only one link is available, DMPO continues to deliver assured application performance by enabling on-demand remediation to mitigate against poor network conditions. The mechanism for on-demand remediation depends on the type of application:

- **Real-Time Application (UDP)**
  For real-time applications, such as voice or video, DMPO sends duplicate packets across a single or two best links in the event of packet loss. Duplicate packets are discarded at the receiving end before forwarding to the application. DMPO also enables on-demand jitter buffering when the WAN links exceed the acceptable jitter threshold.

- **Transactional/Bulk Application (TCP)**
  For Transactional/Bulk applications, such as file transfers, DMPO continues to deliver high throughput over lossy links by using Negative Acknowledgement (NACK). With NACK, the receiving VMware SD-WAN Edge/Gateway signals the sending VMware SD-WAN Edge/Gateway to retransmit the missing packet, therefore protecting the end applications from detecting packet loss and performing TCP retransmission which would lower the effective throughput.

Application Aware Overlay QoS

- **QoS scheduling**
  VMware SD-WAN supports up to 9 traffic classes. As seen in Figure 5.1, a traffic class is a combination of Priority (High, Normal, or Low) and Service Class (Real-Time, Transactional, or Bulk). The resulting 3x3 matrix forms the 9 traffic classes. Each application is mapped into one of the 9 traffic classes. The VMware SD-WAN QoS scheduler ensures all applications in a given class will have a guaranteed minimum aggregate bandwidth during congestion, based on the defined weight, while allowing any application to burst up to the maximum aggregated bandwidth when there is no congestion. The QoS scheduler also ensures fairness among multiple SD-WAN peers, and among multiple flows in a single class. This prevents a single flow or a single peer from using up the maximum allowed bandwidth in the given traffic class.

> **TECH TIP**
>
> The VMware SD-WAN solution comes with Smart Defaults out-of-the-box, which contains pre-defined traffic class assignment for all common application categories, as well as default weights per traffic class. It is recommended to use the Smart Defaults and adjust based on specific application requirements. Each VMware SD-WAN Edge in the network is always aware of available bandwidth of all other VMware SD-WAN Edges to perform smart QoS.

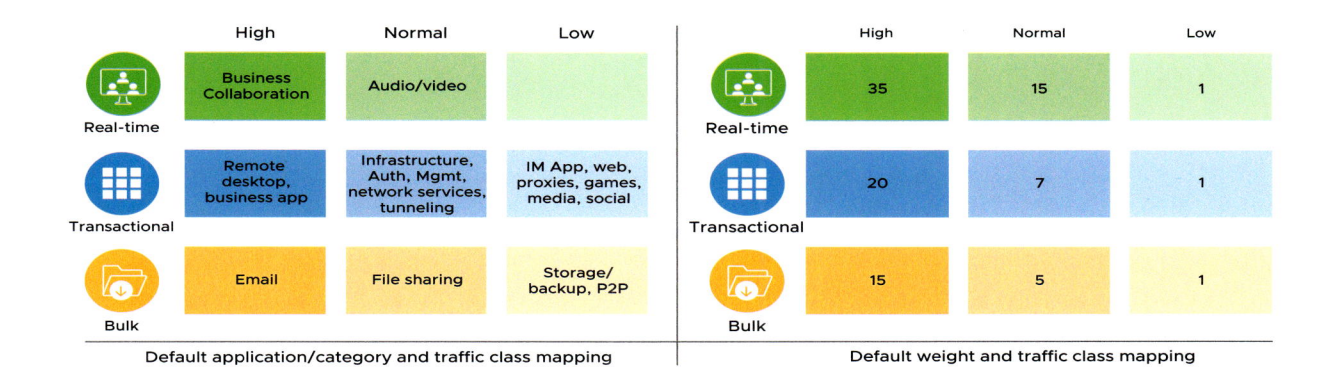

	High	Normal	Low		High	Normal	Low
Real-time	Business Collaboration	Audio/video		Real-time	35	15	1
Transactional	Remote desktop, business app	Infrastructure, Auth, Mgmt, network services, tunneling	IM App, web, proxies, games, media, social	Transactional	20	7	1
Bulk	Email	File sharing	Storage/ backup, P2P	Bulk	15	5	1
	Default application/category and traffic class mapping				Default weight and traffic class mapping		

Figure 5.1: VMware SD-WAN QoS scheme

- **Rate limiting on application or category**
  Rate limiting can be applied per application or application category in both inbound and/or outbound direction. Packets are dropped when traffic rate exceeds the defined rate limit (based on % of link bandwidth) as seen in the `Action` section of Figure 5.2. Navigate to `Configure > Profile > {Profile name} > Business Policy tab > New Rule... > Rate Limit`

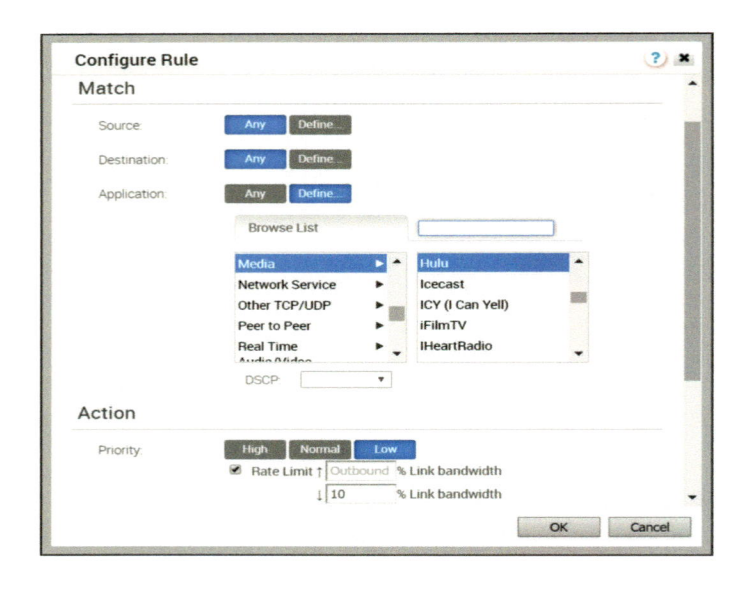

Figure 5.2: Configuration of a Business Policy rule

- **Honoring MPLS CoS**
  To ensure uniform QoS across multiple underlay transports (e.g. internet and MPLS), it is recommended to use a single MPLS CoS for SD-WAN.
  However, VMware SD-WAN can honor MPLS CoS by keeping the Differentiated Service Code Point (DSCP) value in the original payload, while allowing the DSCP value on the tunnel header to be modified or copied from the inner packet (payload). DMPO treats each CoS as a separate WAN link and honors the SLA per class as defined by the MPLS CoS.

Bringing it All Together

The mitigating effect of all mechanisms can be seen from the VMware SD-WAN Orchestrator. When monitoring a VMware SD-WAN Edge on the VMware SD-WAN Orchestrator, there is a QoE tab, which shows something similar to Figure 5.3. Even though the MPLS link was down and the cable internet link was experiencing serious packet loss and mild jitter, the video performance over the SD-WAN Overlay was still good by utilizing all the aforementioned mechanisms.

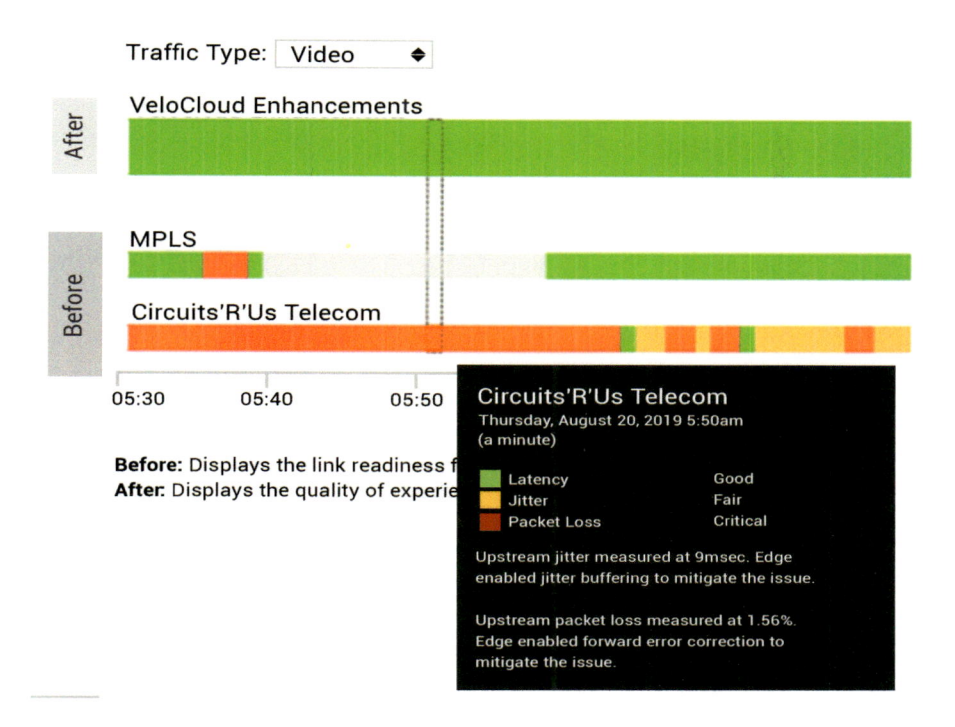

Figure 5.3: Quality of Experience

> Key Takeaways:
>
> - DMPO uses actual traffic to monitor the health of each path in unidirectional fashion. In the absence of user traffic, VCMP probes are sent for monitoring.
> - DMPO provides similar or better performance for applications over internet links using on-demand remediation techniques, even when using a single link.
> - DMPO performs intelligent per packet steering and link aggregation.

Routing

Routing is integral to any organization's network infrastructure. When an organization looks at transforming their network to SD-WAN, one of the first concerns that arises is how SD-WAN devices will integrate with their existing network and how traffic will flow with the mix of a traditional and SD-WAN environment. Most organizations run either BGP, OSPF or static prefixes for their network connectivity, and it becomes critical that the SD-WAN solution supports all of these protocols for successful network migration.

VMware SD-WAN solution fully supports all the standard routing protocols for both underlay (LAN/WAN) and also carries the attributes of routing protocols over the SD-WAN overlay. For example: BGP path attributes such as AS-Path, Local preference, OSPF metrics are preserved and passed over the overlay to other VMware SD-WAN Edges etc. VMware SD-WAN Orchestrator provides a powerful and intuitive UI to configure BGP, OSPF, multicast, so it becomes easier for an organization to make a decision in introducing SD-WAN into their existing network and define policies in order to begin their WAN network transformation.

Figure 5.4 shows the ease and simplicity of the configuration of BGP policies at global profile level which can then easily be pushed down to different sites or regions. Navigate to `Configure > Profile > {Profile name} > Device tab > BGP > Edit`

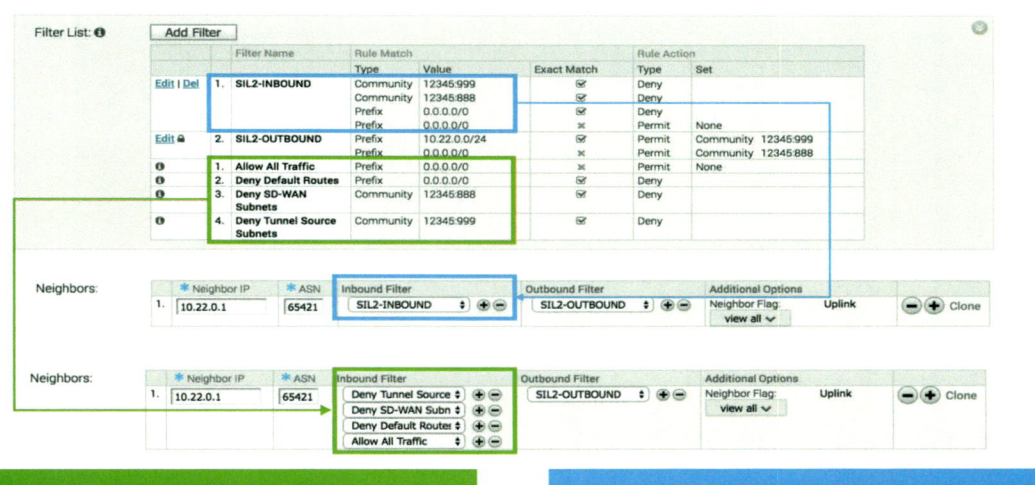

Figure 5.4: BGP configuration example

In addition to traditional routing support, VMware SD-WAN has developed some advanced routing features to make it easier for customers to migrate to SD-WAN and achieve their networking goals. This enables them to see their entire organization's network prefixes and change the routing exit points as per the desired business outcome.

Routing Control Features

Routing control is critical during migration as one of the main challenges is the complexity of dealing with underlay and overlay routing, and how to redistribute routing between traditional MPLS sites and SD-WAN sites.

Prior to SD-WAN, routing across sites was done in the traditional fashion. Each site advertised its own prefixes into the MPLS network, and there was typically only a single next-hop per destination prefix. An exception to this would be for data center prefixes, which could be advertised by multiple data centers. However, route preference for these data center prefixes could be defined based on routing protocol attributes.

With the introduction of SD-WAN, a new routing domain called overlay routing, was introduced on the WAN side. All SD-WAN sites can now reach SD-WAN prefixes via the overlay. If SD-WAN sites are also running a routing protocol with the MPLS provider edge, and redistribution between SD-WAN and MPLS routing domains is not done properly, then there is a potential for sites to become a transit. As shown in the Figure 5.5, multiple sites can now advertise the same prefix resulting in asymmetric routing or sub-optimal routing.

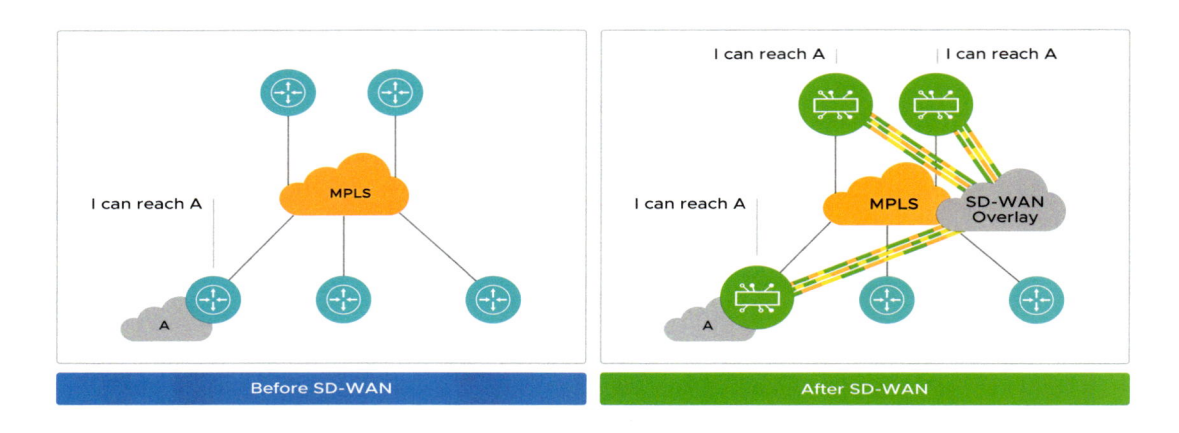

Figure 5.5: The reason to be careful with routing information exchange between overlay and underlay

In order to avoid this asymmetric and complex routing scenarios, VMware SD-WAN solution developed the following features:

BGP uplink neighbor

In the case of a hybrid SD-WAN branch site, there are two sources of routing information: the SD-WAN Edge learns prefixes from—underlay (that is BGP on the MPLS side), and SD-WAN overlay. The mutual redistribution of the prefixes between underlay and SD-WAN overlay is to be avoided, otherwise the branch site will become a transit. Using techniques such as prefix filter to stop the mutual redistribution is cumbersome and error-prone.

VMware SD-WAN solution uses a feature called **Uplink flag** for BGP neighbor that disables mutual redistribution between underlay and SD-WAN overlay. Figure 5.6 shows a hybrid SD-WAN branch setup with uplink flag enabled towards MPLS PE. Enabling this feature prevents the branch site from becoming a transit for traffic from other branches. Note that this will only work when the VMware SD-WAN Edge is inserted in-path i.e. connected to the MPLS CE or PE router.

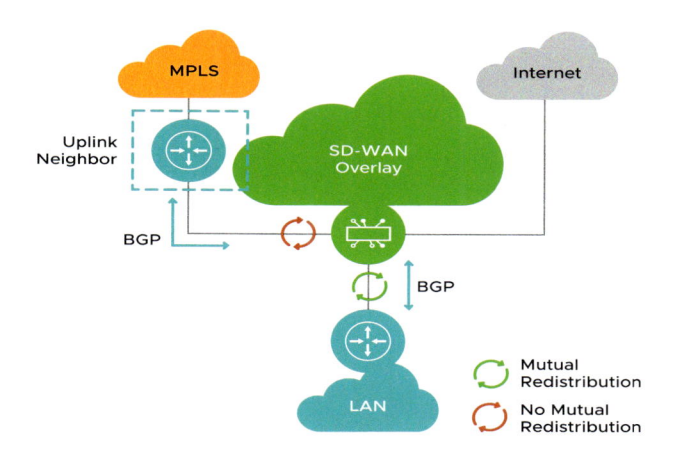

Figure 5.6: BGP Uplink neighbor when the SD-WAN is inserted in-path

Figure 5.7 shows the Neighbor Flag: **Uplink** setting configured under the BGP **Neighbors** setting, which is reached by navigating to `Configure > Profile > {Profile name} > Device tab > BGP > Edit > Additional Options > `**`Neighbor Flag`**:

Figure 5.7: BGP neighbor setting with Neighbor Flag configured to Uplink

BGP uplink community

In the case where the VMware SD-WAN Edge is inserted off-path, it becomes difficult to identify branch LAN prefixes and WAN MPLS prefixes as they are learnt from the same BGP neighbor. In this scenario BGP **Uplink Community** feature can be used. The following figure shows a hybrid SD-WAN branch in an off-path setup with BGP Uplink Community configured. The VMware SD-WAN Edge has a BGP neighbor relationship with the Layer 3 switch, therefore, both LAN prefixes and traditional MPLS prefixes are advertised by the Layer 3 switch. In order to distinguish between the two types of prefixes, all prefixes learned from the MPLS CE router should be tagged with a well-known community, for e.g. 12345:777. This community is then configured as the BGP Uplink Community in the VMware SD-WAN Orchestrator. Prefixes tagged with the BGP Uplink Community are treated the same way as prefixes learned from the BGP Uplink neighbor.

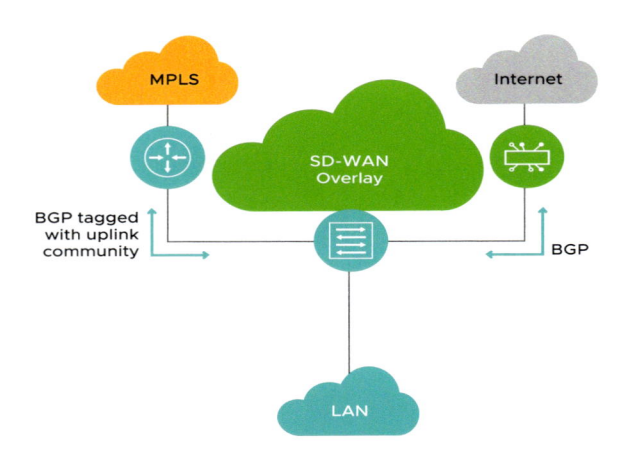

Figure 5.8: BGP Uplink community when VMware SD-WAN Edge is inserted off-path

Figure 5.9 shows the **Uplink Community** configured as 12345:777 in the BGP `Neighbors` configuration which is reached by navigating to `Configure > Edge > {edge name} > Device tab > BGP > Edit > Advanced Settings > Uplink Community`:

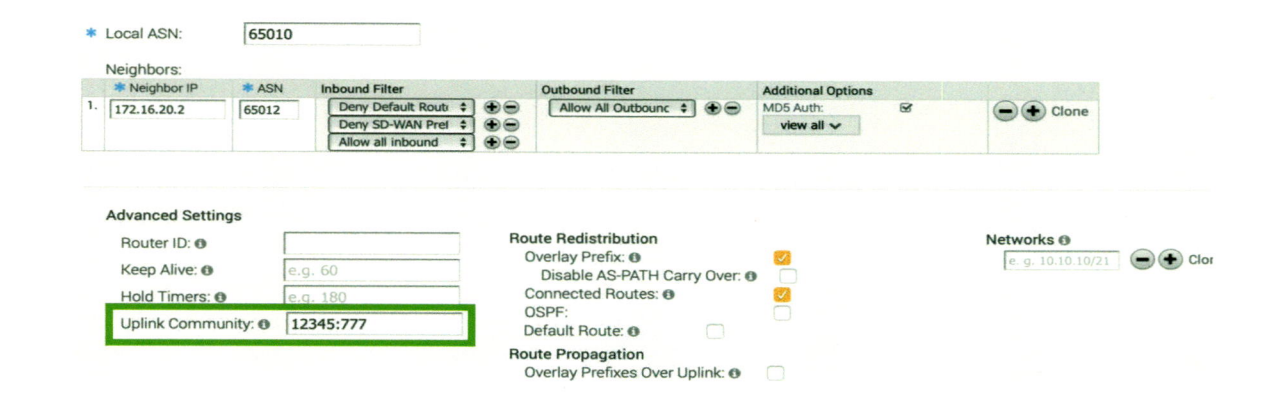

Figure 5.9: BGP neighbor setting with Uplink Community

Overlay Flow Control (OFC)

Traditional network architecture makes it extremely difficult for large organizations to know and control their entire routing table and lacks the visibility to identify route origination on a per site basis. The feature called `Overlay Flow Control` (OFC) was developed for the VMware SD-WAN solution. It gives an organization a single pane of glass for visibility and control into their entire routing table. Figure 5.10 shows the OFC, which is divided into 3 sections, and the following pages provide additional details on each of the 3 sections:

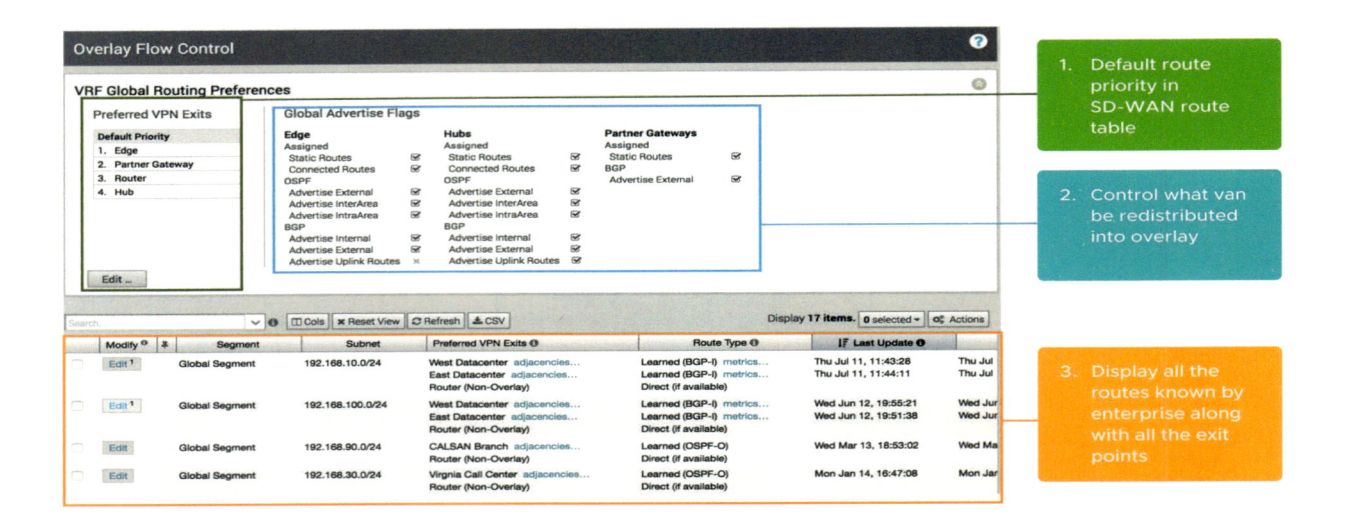

Figure 5.10: Overlay Flow Control

1. Preferred VPN exits: In VMware SD-WAN, prefixes advertised into overlay are categorized into 4 categories:

1. `Edge` prefixes: These are prefixes advertised by other branch or Hub VMware SD-WAN Edges. These prefixes are owned by the respective sites.

2. `Partner Gateway` prefixes: This is only valid for service provider architectures, wherein partner gateways are hosted in the service provider core, such that the partner gateway peers with PEs and advertises learnt prefixes from PEs into the SD-WAN overlay.

3. `Router` prefixes: Any prefixes learnt by VMware SD-WAN Edges by BGP (eBGP/iBGP) or OSPF.

4. `Hub` prefixes: These are special category prefixes and have the lowest default preference. Branch VMware SD-WAN Edges will only use these prefixes to send traffic if there are no other Edge or Router prefixes available. Classification of a hub prefix is discussed in the following scenarios:

   - The Hub learns a prefix via an uplink BGP neighbor or learns a prefix tagged with an uplink community and advertises it into SD-

WAN overlay. When branches receive this prefix from the SD-WAN overlay they will be marked as hub prefixes.

- External prefixes learnt via OSPF (E2/E1) at hubs.

Figure 5.11 below provides an example to explain the Preferred VPN Exits that align with list above:

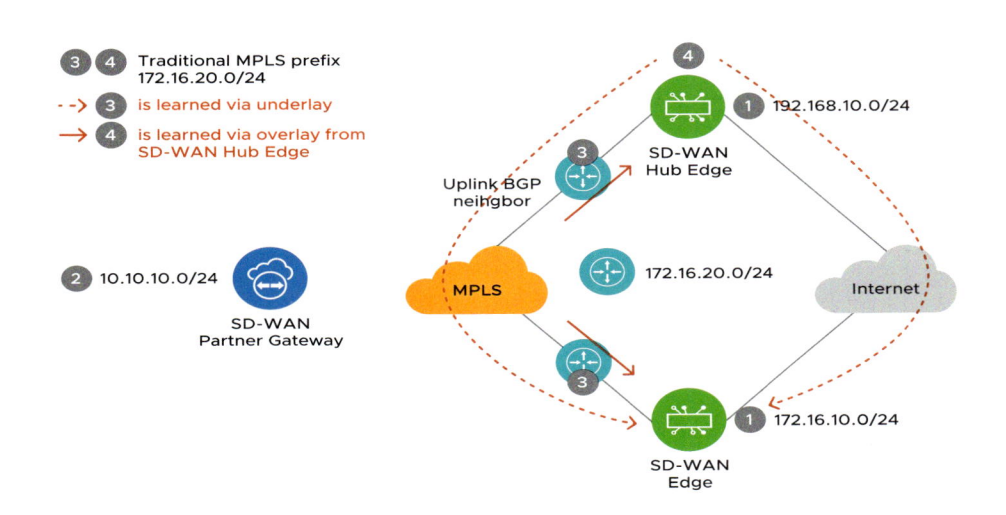

Figure 5.11: OFC Preferred VPN Exits example

2. Global advertise flags: These flags are strictly related to what type of prefixes are allowed to be redistributed into the SD-WAN overlay. Flags are divided into three different site types.

- Edge flags: These flags dictate what type of prefixes branch VMware SD-WAN Edges can be advertised from underlay to overlay.

- Hub flags: These flags dictate what type of prefixes hub VMware SD-WAN Edges (SD-WAN Edges classified as Hubs in profiles) can advertise from underlay to overlay.

- Partner gateway flags: These flags only apply to partner gateways hosted by service providers and dictate what prefixes learnt from the PE that partner gateways can advertise into the overlay.

3. Routing table: This table lists the entire organization's overlay routing table and the preferred VPN exit points highlight which sites own or advertise which prefixes and use which routing protocol.

- For example, in Figure 5.10: 192.168.10.0/24 & 192.168.100.0/24 are originated by West datacenter in the global segment. And East datacenter is acting as backup for these prefixes. Both Hubs learnt these prefixes via iBGP and time stamps when they last learnt it are also displayed. All branch sites will send traffic over the overlay tunnels to West datacenter and use East datacenter as backup.
- Similarly, 192.168.90.0/24 is advertised into overlay only by CALSAN Branch. CALSAN Branch learnt this prefix via OSPF from its LAN side.
- If a SD-WAN Branch Edge receives a prefix from two different hubs with the same metric then the higher order Hub defined in the profile will be the preferred VPN exit for that SD-WAN Branch Edge.

Key Takeaways:

1. VMware SD-WAN allows for full configuration of BGP, OSPF, and Multicast by powerful and intuitive UI.
2. VMware SD-WAN features such as BGP uplink neighbor and BGP uplink community simplify route control during migration.
3. Overlay Flow Control (OFC) provides a holistic view of an organization's entire route table along with information of the sites originating the prefixes.

Business Policy

In order to understand how applications are handled within the VMware SD-WAN solution, it is important to understand how packets are processed within a VMware SD-WAN Edge. Figure 5.12 illustrates the order of operation.

1. **Route Table Lookup**: The first operation is a route table lookup where the decision is made to either send it to the `Overlay` (`Cloud VPN`) or `Underlay` and whether traffic is bound to the internet. Matching the default route of type `Cloud Gateway` also enables the business policy to change the internet breakout behavior (e.g. send traffic via the VMware SD-WAN Gateway)

2. **Application Recognition:** The deep application recognition (DAR) engine identifies the kind of traffic as the second order of operation.

3. **Business Policy:** The business policy classifies the applications into different QoS classes and applies the policies such as service insertion, rate limiting, and link steering. Link steering and service insertion are only available for route types Cloud VPN and Cloud Gateway.

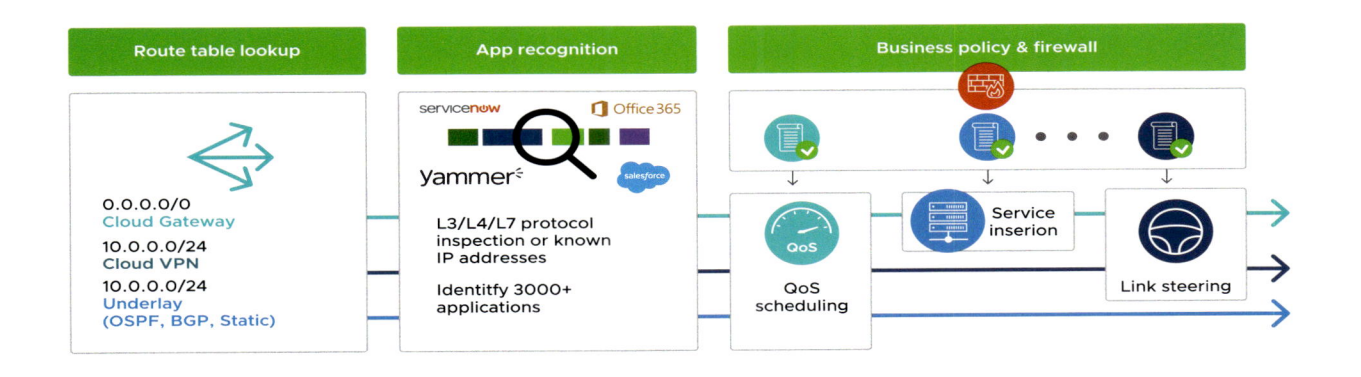

Figure 5.12: Order of operation

Application Performance, Routing, and Cloud Access

> **TECH TIP**
>
> In contrast to traditional networks, it is best practice to not learn/advertise a default route into the overlay as this would overrule the Cloud Gateway Route.

| 0.0.0.0 | Global | 0.0.0.0 | Cloud | 0 | TRUE | Cloud Gateway |
| 0.0.0.0 | Global | 0.0.0.0 | Cloud | 5 | TRUE | GE3 |

Figure 5.13: Route table snippet with Cloud Gateway route

Please refer to the Chapter 6: *Inspecting User Traffic* for further details on using business policy for managing internet access.

General Behavior of Business Policies

For creating or changing a business policy the following needs to be kept in mind:

- Business policies are processed sequentially.
- New rules will be put at the top, so re-arranging by drag and drop might be required.
- Changes to the business policy that alters the traffic flow direction are not applied to already existing network flows.

Business Policy Rule Creation

VMware SD-WAN business policies follow the well-known concept of Match and Action. As seen in Figure 5.14, combining the routing table and app recognition with optional manual configuration gives a powerful tool to treat traffic granularly.

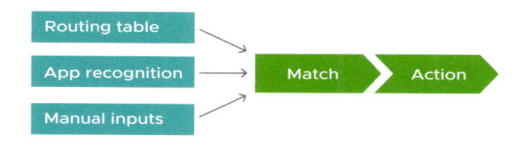

Figure 5.14: Business policy—Match & Action

VMware SD-WAN also provides out-of-the-box business policy rules that are tuned to different application types and their QoS, prioritization, and scheduling needs. They are referred to as **Smart Defaults** and they enable a customer to turn up a VMware SD-WAN deployment with limiting the need to perform any manual configuration. Customers can immediately use and benefit from dynamic packet steering and application-aware prioritization without having to define policies.

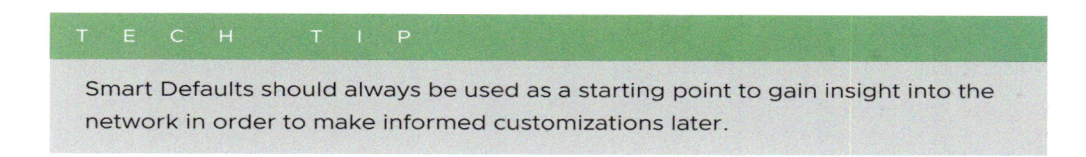

> **TECH TIP**
>
> Smart Defaults should always be used as a starting point to gain insight into the network in order to make informed customizations later.

Creating a Business Policy Rule

Business policies can be created on a per-profile and per-device level, enabling a flexible rollout throughout the network.

On the VMware SD-WAN Orchestrator, Navigate to `Configure > Profile > {Profile name} > Business Policy tab > `**`New Rule:`**

Next to giving the rule a name, the Match and Action parts of the business policy are straightforward to configure as seen in Figure 5.15.

Application Performance, Routing, and Cloud Access

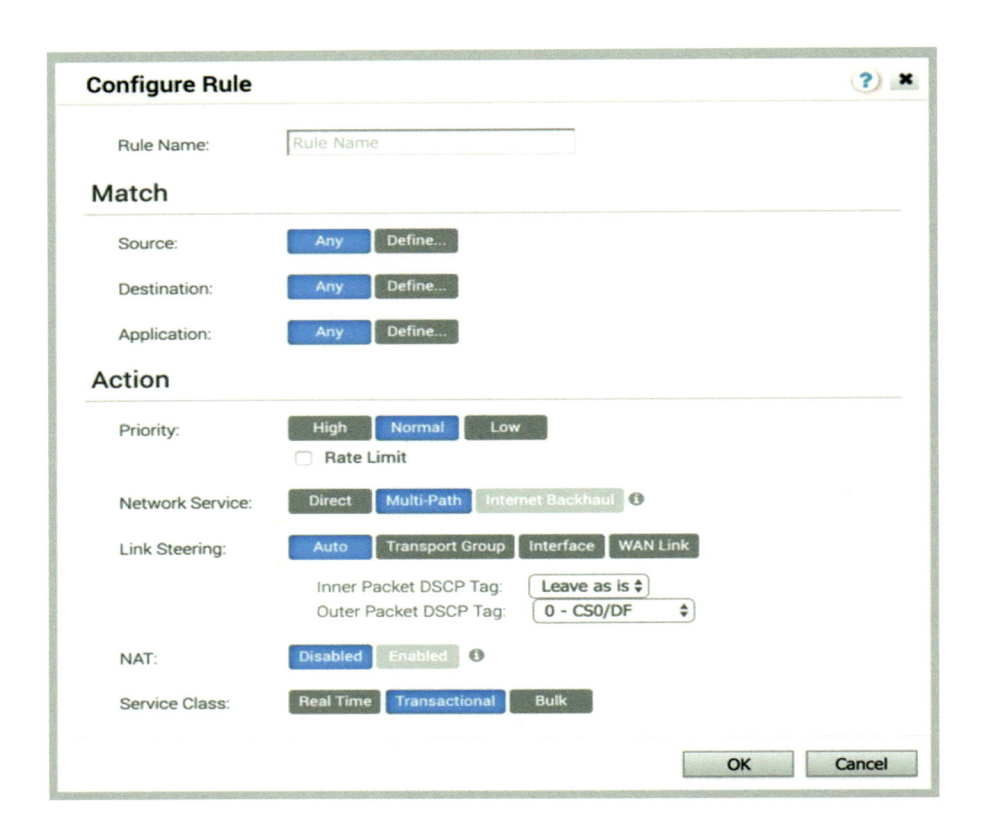

Figure 5.15: Configuration of a business policy rule

Matching packets for further processing

As seen in Figure 5.16, the `Match` section contains the application recognition and additional filtering based on `Source` and `Destination` to precisely filter the traffic to be processed in the `Action` section. Navigate to `Configure > Profile > {Profile name} > Business Policy tab > New Rule:`

`Source` and/or `Destinations` can be matched on a variety of characteristics such as VLAN numbers, IP addresses, subnets/wildcard mask, port numbers, hostnames, protocols. Based on the DAR engine, individual applications or groups of applications can be selected to be matched. `Application` maps, which can be applied by an operator on the VMware SD-WAN Orchestrator, allow organizations to extend the list of recognized applications with their own specific applications if needed.

> **TECH TIP**
>
> **Beware:** Configuring secured internet breakout requires the destination to be of type **Internet**. This matches the default Cloud Gateway route and allows flexible backhauling or handoff to Cloud Access Security Brokers (CASB)

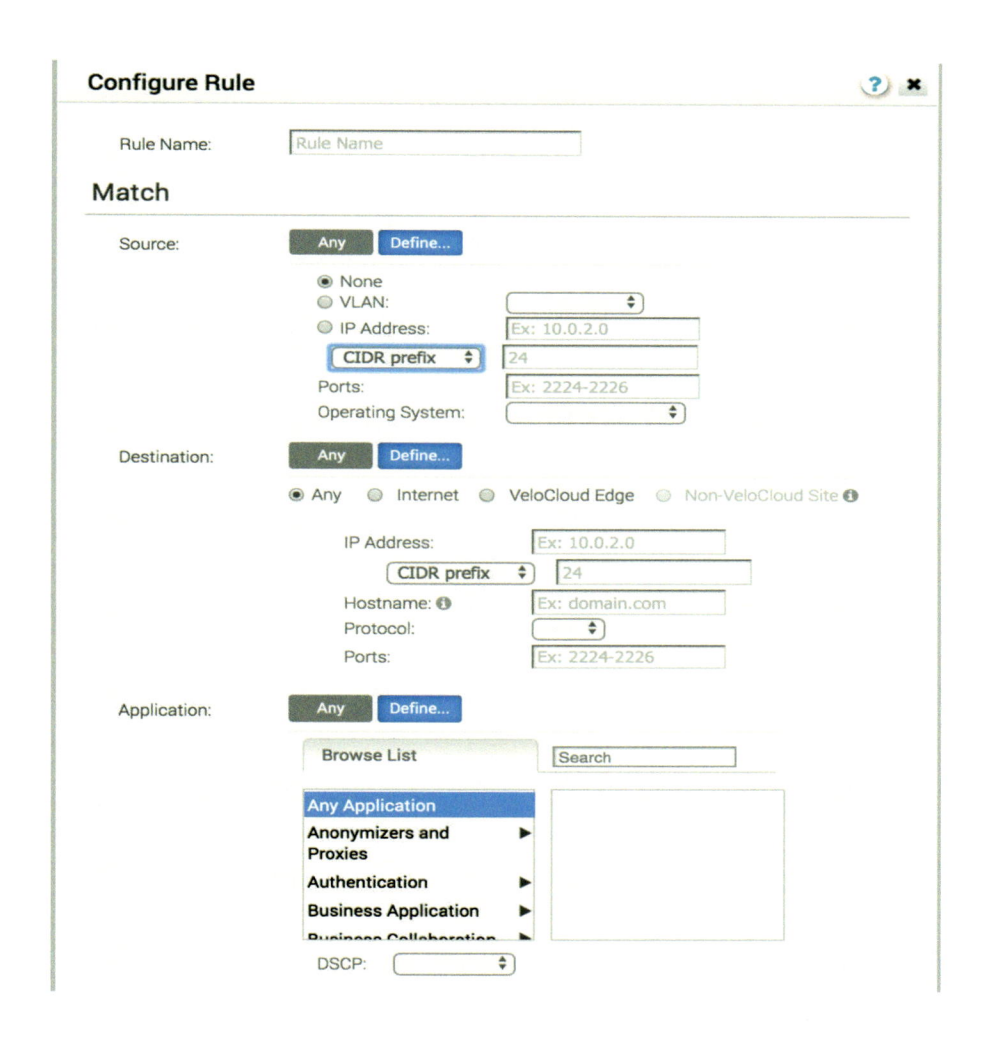

Figure 5.16: Business policy rule—Match section breakdown

Define the actions for matched traffic

Within the `Action` section, all further treatment of the interesting traffic is defined as seen in Figure 5.17.

- `Priority` and `Service Class` are an integral part of the VMware SD-WAN 3x3 QoS settings, defining not only the queue but also the available On-Demand Remediation executed in case of network impairment detection.

- `Network Service` is used for determining the internet breakout. This is further explained in Chapter 6: *Security* section of this book.

- `Link Steering` is used to configure steering matched traffic to certain transport technologies up to explicit interfaces. It also contains the configuration for DSCP marking and remarking.

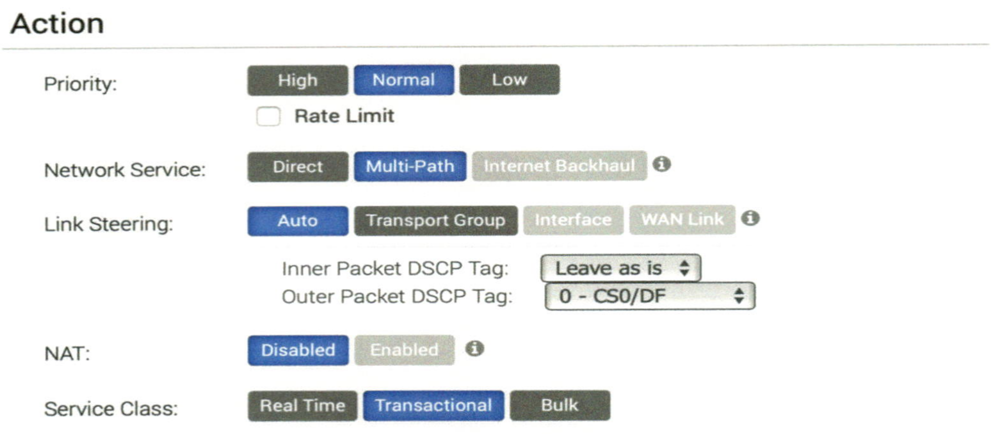

Figure 5.17: Business policy rule—Action section breakdown

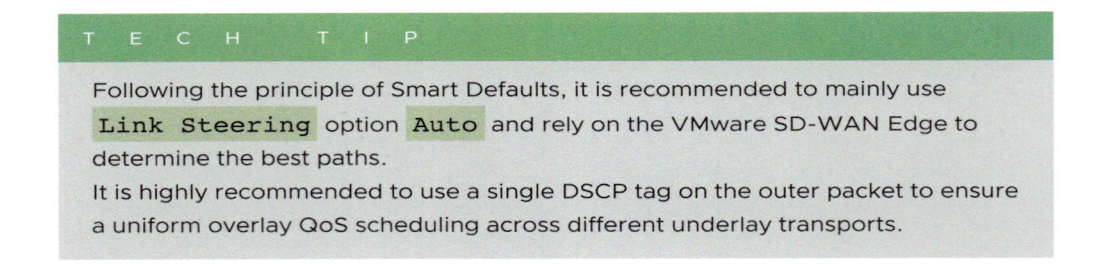

T E C H   T I P

Following the principle of Smart Defaults, it is recommended to mainly use `Link Steering` option `Auto` and rely on the VMware SD-WAN Edge to determine the best paths.

It is highly recommended to use a single DSCP tag on the outer packet to ensure a uniform overlay QoS scheduling across different underlay transports.

Business Policy Example

Figure 5.18 gives a real-world example where a hybrid SD-WAN branch with an additional LTE is deployed to achieve maximum application performance and reliability. A fine tuning of the available Smart Defaults might be helpful in this case to influence the utilization of the branch's connectivity, using the business policy framework.

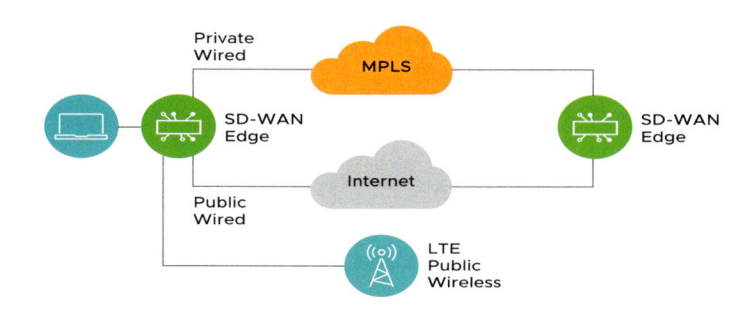

Figure 5.18: Sample network topology for business policy example

Looking at the corresponding **Rule** numbers in Figure 5.19 for this network topology, the network engineer plans to fulfill the following requirements. Navigate to `Configure > Profile > {Profile name} > `**`Business Policy tab:`**

1. Backup traffic shall use the wired internet connection for cheap bandwidth as long as possible.

2. Collaboration shall use the MPLS connection up to a degradation of 2% Packet Loss or jitter higher than 15ms.

3. Client OS update traffic must only use internet and is not allowed to consume more than 10% of the available BW and is not allowed to failover to the public wireless connectivity available via the LTE transport.

The rest of the applications are covered by Smart Defaults and don't need any special adjustments.

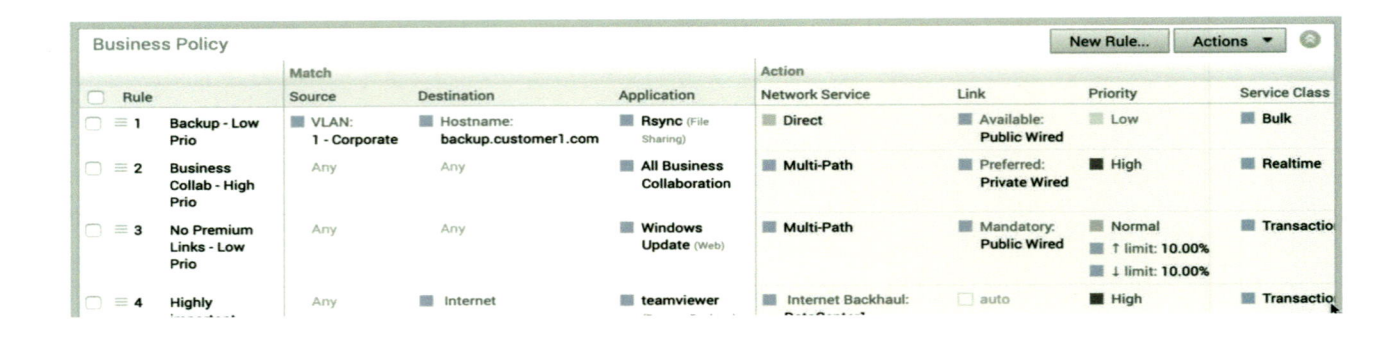

		Match			Action			
☐	Rule	Source	Destination	Application	Network Service	Link	Priority	Service Class
☐ ≡ 1	Backup - Low Prio	■ VLAN: 1 - Corporate	■ Hostname: backup.customer1.com	■ Rsync (File Sharing)	■ Direct	■ Available: Public Wired	▫ Low	■ Bulk
☐ ≡ 2	Business Collab - High Prio	Any	Any	■ All Business Collaboration	■ Multi-Path	■ Preferred: Private Wired	■ High	■ Realtime
☐ ≡ 3	No Premium Links - Low Prio	Any	Any	■ Windows Update (Web)	■ Multi-Path	■ Mandatory: Public Wired	■ Normal ■ ↑ limit: 10.00% ■ ↓ limit: 10.00%	■ Transactio
☐ ≡ 4	Highly	Any	■ Internet	■ teamviewer	■ Internet Backhaul:	☐ auto	■ High	■ Transactio

Figure 5.19: Business policy for sample network topology requirements

Appling this to the branch profile makes a consistent app performance routing available to all branches without the need to configure each branch individually. In order to make rule no. 2 work, the option for **Error Correct Before Steering** must be used as shown in Figure 5.20. Navigate to **Configure > Profile > {Profile name} > Business Policy tab > {Rule name} > Preferred & Error Correct Before Steering**

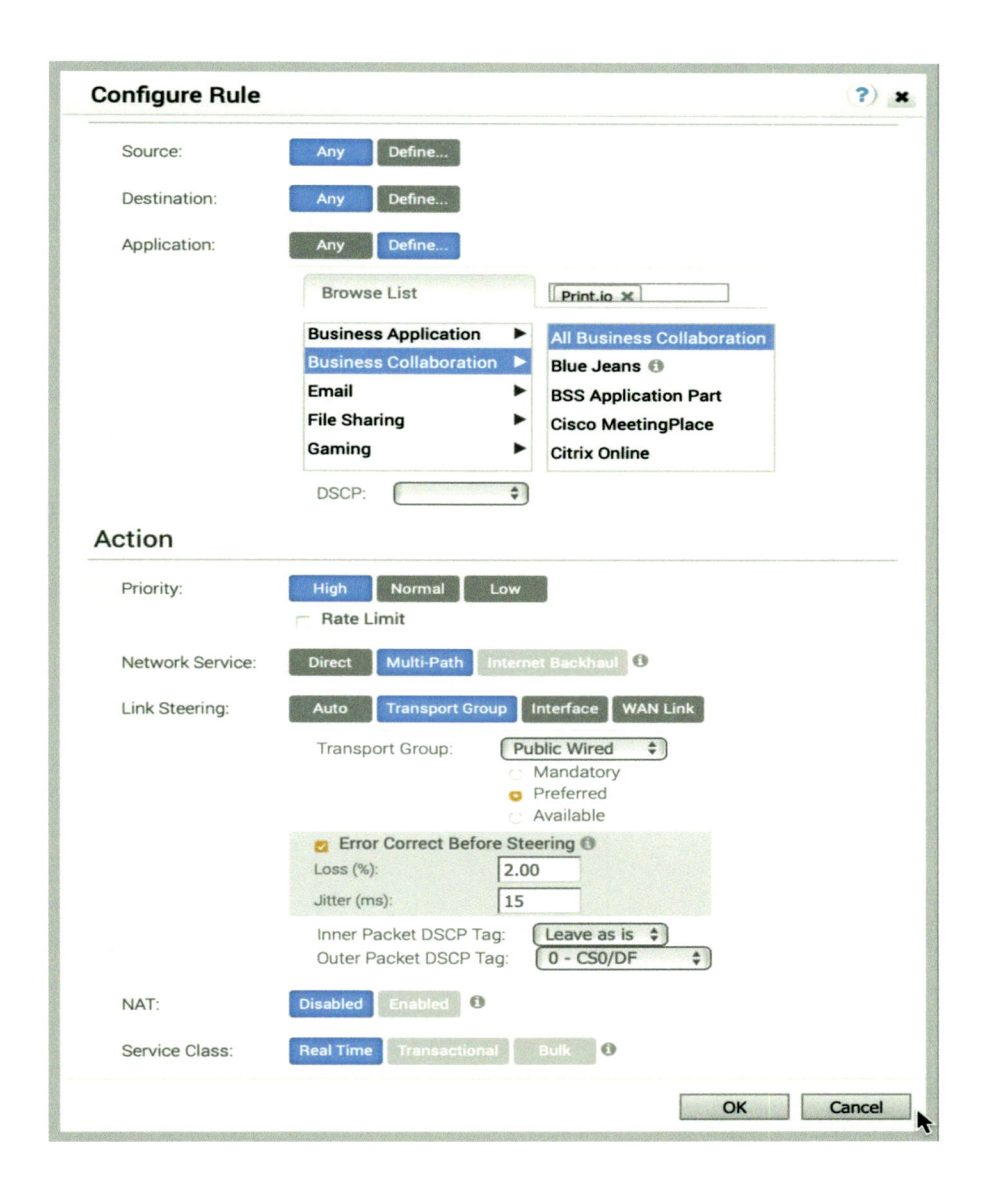

Figure 5.20: Business policy rule including Error Correct Before Steering option

As seen in the flow example in Figure 5.21, the policy setting will be automatically mirrored on the other edge involved in the flow automatically, so that also the return traffic is treated accordingly.

| | | Global | TCP | 9080 | 38742 | ssl | | Loadbalance | Branch to Branch Direct | Refer Policy on Peer Edge Device |

Figure 5.21: Flow example on the receiving end of an adjusted business policy

> **Key Takeaways:**
>
> - Business policies provide a simple yet comprehensive tool to an organization to configure their QoS, security, steering, and service insertion policies.
> - Understanding the order of operation is an important pre-requisite before modifying Business Policy Rules.
> - VMware SD-WAN Smart Defaults provides a powerful and simple way of ensuring application performance.
> - Business policy configuration is synced between VMware SD-WAN Edges and between VMware SD-WAN Edge and VMware SD-WAN Gateways.

Optimized Cloud Access

As more and more workloads migrate to the public cloud, the quality, constrained bandwidth and cost of private links are becoming a big concern to enterprise IT, as this traffic typically needs to be backhauled from the branch to the data center. IT needs to look for scalable, secure, and optimized access to Infrastructure-as-a-Service (IaaS) and Software-as-a-Service (SaaS) from remote branches. Traditional VPN technologies offered by vendors to access the public cloud are complex, time-consuming, and often lack enterprise-grade performance and security.

This chapter explains the options VMware SD-WAN provides for connecting organization branches to the public cloud.

SaaS Access via VMware SD-WAN Hosted Gateway

Organizations can choose to send SaaS traffic via hosted multi-tenant VMware SD-WAN Gateways, thereby taking advantage of last-mile optimization, link aggregation, and on-demand remediation capabilities of DMPO.

As seen in the Figure 5.22 below, once traffic from the SD-WAN Edge reaches the VMware SD-WAN Gateway, it will get port address translation (PAT) via VMware SD-WAN Gateway's public IP address.

Figure 5.22: SaaS access via hosted VMWare SD-WAN Gateways

The following Figure 5.23 shows the configuration to send SaaS traffic via Hosted VMware SD-WAN Gateway. Navigate to `Profile > {Profile name} > Business Policy tab > New Rule {Rule name} > Action > Network Service > **Multi-Path**`

Figure 5.23: Business policy rule multi-path

IaaS Access via SD-WAN Hosted Gateway

In this option, the public cloud VPN gateway is considered a Non-VeloCloud Site (NVS). Refer to the Figure 5.24 for the topology.

The VMware SD-WAN Edge will establish an SD-WAN Overlay tunnel with its assigned VMware SD-WAN Gateway. That VMWare SD-WAN Gateway in turn will form a standards-based IPsec tunnel to the public cloud VPN gateway. The selection of the VMware SD-WAN Gateway is automatically performed by the VMWare SD-WAN Orchestrator. This selection occurs on the basis of the proximity between the available VMware SD-WAN Gateways and the public cloud VPN gateway.

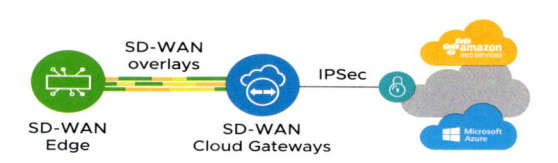

Figure 5.24: IaaS access via SD-WAN Hosted Gateway

The advantage of this option is that there isn't a need for multiple public cloud VPN gateways, nor does it require changes to the enterprise data center. In addition, there will only be one IPSec tunnel required from the public cloud VPN gateway, compared with the creation of numerous tunnels in the event the IPSec is initiated from each branch. This simplifies the design and achieves scale at the same time.

> **TECH TIP**
>
> To achieve full redundancy, configure two VPN Gateways (aka VPN connections to AWS VPC or Azure vNet), and select Redundant VeloCloud Cloud VPN to assign two VMware SD-WAN Gateways for VPN terminations.

The configuration of the Non-VeloCloud Site (NVS) is as simple as two steps, Figure 5.25 is a sample screenshot showing the required configuration. Navigate to `Configure > Network Services > Non-VeloCloud Site > New...`

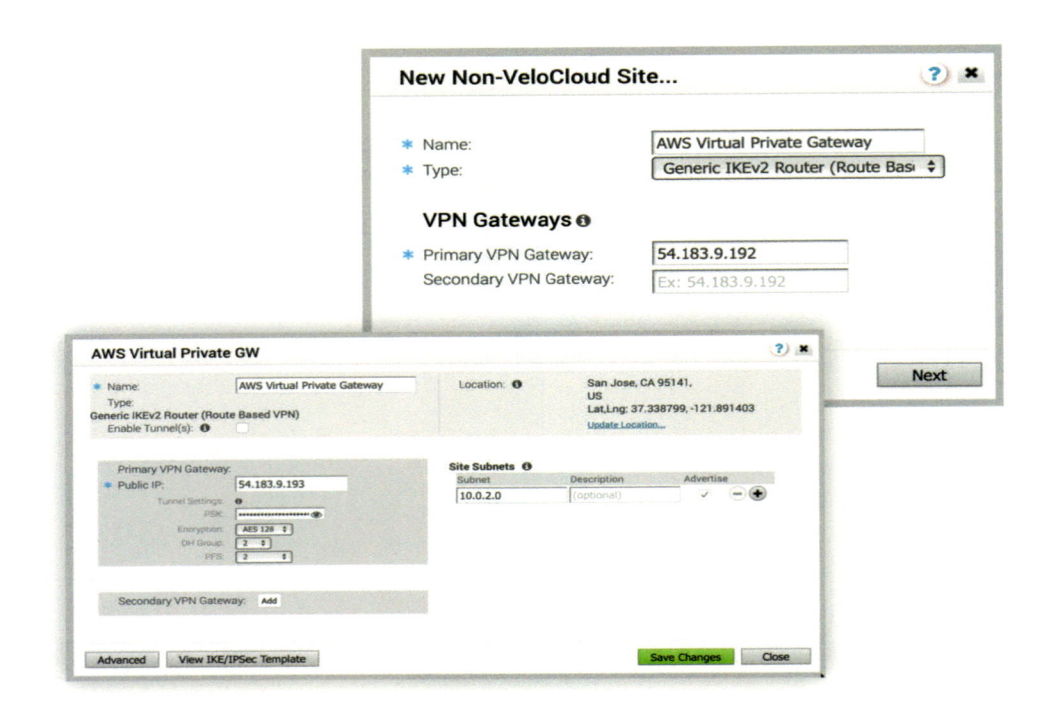

Figure 5.25: Configuration of non-VeloCloud site (NVS)

IaaS Access via VMware SD-WAN Edge on Public Cloud

The virtual VMware SD-WAN Edge is also available in AWS/Azure marketplaces. By deploying the virtual edge in the AWS VPC or Azure vNET rather than leveraging the public cloud VPN gateway, customers' public cloud instances immediately become another VMware SD-WAN site. This means that the connectivity between the public cloud and SD-WAN sites can enjoy all the benefits of the VMware SD-WAN solution as well. This includes end-to-end DMPO optimization benefits and consistent policies extended to the public cloud. A typical scenario for this type of deployment is the VMware SD-WAN Virtual Edge having a single internet connection as depicted in Figure 5.26. For customers who have high multi-Gigabit throughput needs, multiple virtual VMware SD-WAN Edges can be clustered with resource-aware clustering technology offered by the VMware SD-WAN solution (see Chapter 4: *Site Topology and Redundancy*.

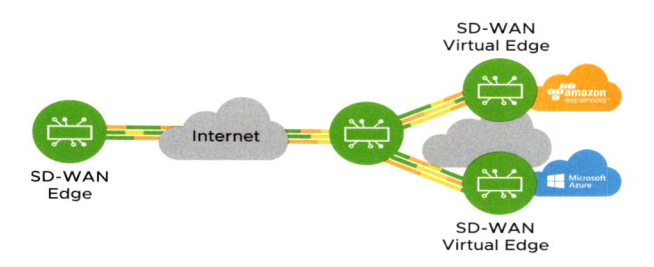

Figure 5.26: IaaS access via VMware SD-WAN virtual Edge deployed in AWS VPC or Azure vNet.

In a situation such as the public cloud having a direct connection to the customer's backbone network, the virtual edge can utilize both internet and private circuit as seen in Figure 5.27.

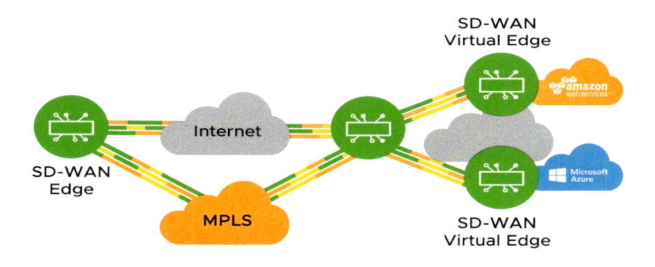

Figure 5.27: End-to-end DMPO with dual circuit to VMware SD-WAN Virtual Edge in the cloud

Templates for deploying the virtual edge, such as AWS's CloudFormation template, are also available to facilitate ease of deployment of the virtual edge in public cloud environments.

Service provider with dedicated cloud connection

Often the service provider has already established a dedicated cloud connection to common SaaS / IaaS. VMware SD-WAN Gateways to facilitate customer connections to the public cloud. In this scenario, the VMware SD-WAN Gateway assumes the role of a partner gateway. Operating in this role, the partner gateway will have two network interfaces. One interface that will connect to the internet, and the other interface will connect to the service provider's Provider Edge router that already has a dedicated connection to public cloud. This option enables the service provider to offer the internet as a connectivity option for its customer to connect to the VMware SD-WAN Gateway and use it as a means to access the public cloud. With this topology, the internet's last mile (between customer and VMware SD-WAN Gateway) can enjoy the benefits of DMPO optimization. Figure 5.28 shows the topology of this option:

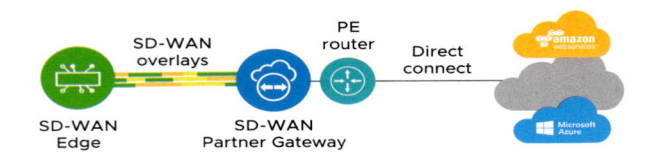

Figure 5.28: Service provider with direct connection

Connecting to Microsoft Azure over vWAN

Microsoft offers Microsoft Azure Virtual WAN (vWAN) globally. This network service provides optimized connections over the WAN to resources on Azure. Microsoft Azure vWAN allows customers to easily access the Azure global network that provides very low latency and optimal routing. By utilizing VMware SD-WAN to connect to vWAN, not only is the last mile improved via DMPO enhancements, network visibility is also improved.

The methods of connecting to the Microsoft Azure vWAN are similar to methods described earlier in *IaaS Access via SD-WAN Hosted Gateway*. Figure 5.29 shows the topology when integrating VMware SD-WAN with Microsoft Azure vWAN.

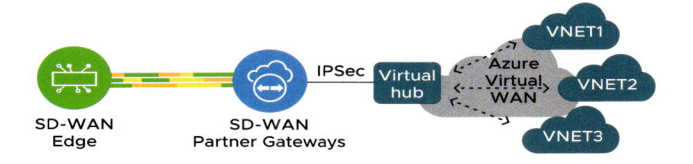

Figure 5.29: VMWare SD-WAN connectivity to Azure vWAN

In the vWAN use case, the NVS now becomes the vWAN hub. This means the VMWare SD-WAN Gateway will form a standards-based IPSec tunnel to the vWAN Hub. There is no change for the connectivity between VMware SD-WAN Edge and VMware SD-WAN Gateway. The connectivity between the VMware SD-WAN Edge and VMware SD-WAN Gateway is an SD-WAN Overlay tunnel, which provides secure connections and also enhancements by DMPO. Figure 5.30 shows the options specific to vWAN, such as `Subscription` and `Virtual Hub`. Navigate to `Configure > Network Services > Non-VeloCloud Site > New > Microsoft Azure Virtual Hub`:

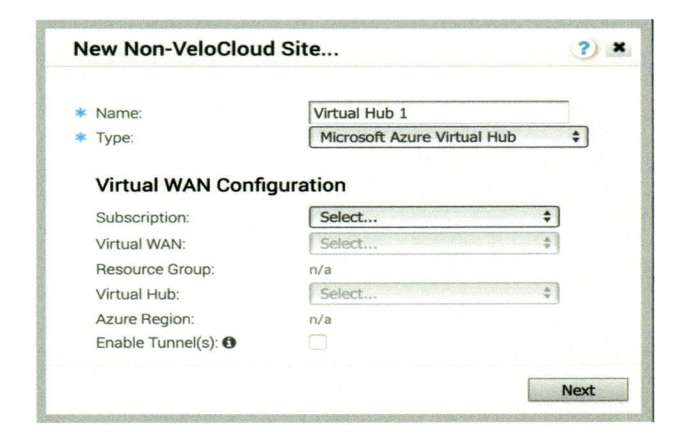

Figure 5.30: Creation of Microsoft Azure vWAN Hub as a non-VeloCloud site

Application Performance, Routing, and Cloud Access

Key Takeaways

- Cloud-hosted VMware SD-WAN Gateways provides multiple flexible and scalable solutions to connect to SaaS and IaaS without the need to backhaul all branch traffic to the organization's data center.
- Service providers can leverage VMware SD-WAN Partner Gateways to offer its customers optimized cloud access.
- The VMware SD-WAN solution provides fully automated connectivity to Azure vWAN.

Rodney had been pessimistic about using the VMware SD-WAN default business profiles to protect Allofu's critical business applications during the North Dakota pilot. Barbara from Rodney's team convinced him to follow through, once again using the question "What do we have to lose?" Barbara has a real knack for seeing when the team has gotten sidetracked and the group realizes that she's right; it's finally "Go Time!"

Even though it throws a wrench into his dinner plans, Rodney elects to prove the zero-touch-provisioning activation to see if it lives up to VMware's claims. Rodney asks Leia to ship the VMware SD-WAN Edge to the North Dakota Branch and proceeds to complete the activation steps via the VMware SD-WAN Orchestrator. The VMware SD-WAN Edge arrives the next business day and Rodney walks Matty, the Branch Manager, through the physical activation of the appliance. Rodney is impressed as he watches the VMware SD-WAN Edge come online and he verifies connectivity. Rodney anxiously waits as Matty verifies that his applications work as expected. Matty, of course, is unimpressed and impatient to leave for the evening of his Beer League hockey game. Rodney thinks who in their right mind goes ice skating when it's already -20 degrees outside? Rodney then realizes he has never seen -20 degree's and doesn't want to. Zero-touch-provisioning is a true lifesaver! Rodney sends Alvina an update of the successful site activation and initial testing of the applications. Now it's time to wait and watch as once the users arrive the real test begins. Now that the pilot has officially kicked off, Rodney needs to dig into the configuration required to configure the business policies for Allofu's applications. Alvina reads Rodney's email and is optimistic that budget relief could be on the horizon if the SD-WAN pilot goes well.

Even though the Allofu branch only has a single link, the users have been eerily, yet pleasantly, silent since activation of the VMware SD-WAN Edge several days ago. In fact, Rodney has needed to double check that the team in North Dakota wasn't on holiday as he cannot recall a period this long without complaints from those problem children. Rodney has begun to explore the simple and intuitive graphics provided showing the link degradation to his provider Circuits 'R' Us, who now have some explaining to do!

Alvina is encouraged by the early progress of the pilot. Alvina has checked in with her sales leaders and sure enough, they are happier than usual. Alvina called Matty to check in on how the pilot was going from his perspective. This was a call she was dreading; he is well known for being generally sour. Matty was in his typical dark mood but had found entirely new issues to complain about and for the first time in her recollection, none of them were about network or application performance. Alvina inquired specifically about VOIP calls and Matty said everything was "Good." Good is the equivalent of amazing when coming from Matty and Alvina takes this as ringing endorsement.

From speaking with Rodney, Alvina learns that there are indeed some link level issues that he is investigating with Circuits 'R' Us and Rodney explains that he previously couldn't easily see this information. Alvina requests that Rodney also pilot a few additional locations to further validate internet-based connectivity and even an LTE option for Henrik, the creative genius living in a campground somewhere that sounded truly ghastly. Alvina reminds herself that the creative types are always a bit "eccentric," but man was he ever good at his job!

6. Security

Rodney is getting more and more excited about the VMware SD-WAN possibilities and walks into his weekly meeting with Alvina. To his utter horror, Hans, Allofu's security architect, is sitting in Alvina's office and they are evidently waiting for him. The look on Hans' face clearly indicates that his demeanor remains as "cheerful" as ever...this will be a "fun" meeting.

Alvina urges Rodney to sit down and Hans immediately starts interrogating Rodney about the proposed SD-WAN solution. Hans is extremely concerned about the security of a cloud-hosted solution and the possible security implications for allowing users to reach the internet directly without first going back to the data center where his security appliances are located.

Alvina is concerned as she flashes back to those recurring nightmares she's been having. Surely Allofu is not the first customer to raise these concerns. How are customers addressing these concerns?

Rodney isn't exactly clueless when it comes to security, but it isn't his forte either. Rodney needs to quickly understand and figure out how to address the concerns raised by Hans. Luckily, he is off to goat yoga shortly and he can ask his buddy Yves who mentioned that his organization had moved to cloud-based security.

It's also time to have Leia come back in and explain what VMware has seen its customers doing to address these concerns.

When selecting an SD-WAN solution, security is a key concern for any organization. There are three primary considerations in relation to network security:

1. Security of the overall solution
2. Security of the components
3. Inspection of user traffic

In the following chapters, we will discuss each of these considerations. In addition, network segmentation and best practices for how to secure SD-WAN deployment will be discussed.

Security—VMware SD-WAN

Security is fundamental to the VMware SD-WAN solution which is built on an architecture that ensures secure communication between the management, control, and data planes. As shown in Figure 6.1, the management plane consists of the VMware SD-WAN Orchestrator and the control plane is comprised of the VMware SD-WAN Controller (a function of the Gateway). In the hosted deployment, the VMware SD-WAN Gateway/Controller has a dual function in that it can optionally participate in the data plane when the VMware SD-WAN Gateway is set to process internet-bound traffic to take advantage of DMPO capabilities. In an on-premises deployment, the controller operates purely as a control plane component.

Figure 6.1 shows the methods used to secure communication between control, data, and management planes.

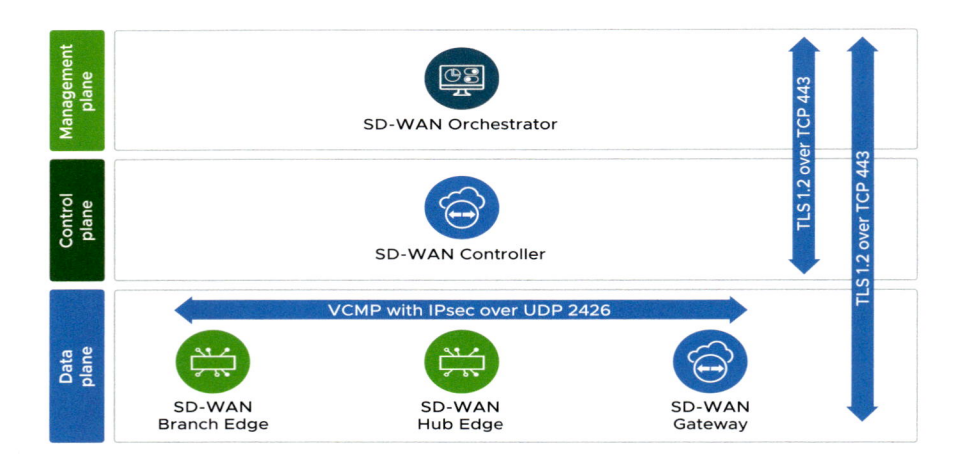

Figure 6.1: Secure communication between VMware SD-WAN components

Security between the management and data planes

When a new VMware SD-WAN Edge or VMware SD-WAN Gateway is first provisioned, an **activation key** is generated by the VMware SD-WAN Orchestrator. This activation key is used by the new device to authenticate and download configurations and policies from the VMware SD-WAN Orchestrator. Along with the configuration and policy download, an **identification token** is issued to the VMware SD-WAN Edge/Gateway at time of activation. The identification token allows the VMware SD-WAN Orchestrator to uniquely identify all the devices in its management domain. The process described here happens seamlessly during the zero-touch provisioning.

VMware SD-WAN Edges/Gateways establish a TLS 1.2 encrypted session to the VMware SD-WAN Orchestrator to receive device certificate, configuration and policy updates, and to upload statistics and event logs. The VMware SD-WAN Orchestrator's public CA signed certificate is used to secure the transport. Once the TLS 1.2 connection is established, the VMware SD-WAN Orchestrator performs an identification match using the identification token sent by the VMware SD-WAN Edge/Gateway against the one stored in the orchestrator database corresponding to that particular device. In case of an identification mismatch, the VMware SD-WAN Orchestrator will deactivate that device.

Security between data plane and control plane components

Traffic between VMware SD-WAN Edges and from VMware SD-WAN Edges to VMware SD-WAN Gateways uses VCMP tunneling over UDP port 2426, secured with IPSec. One exception to this is for traffic destined to the public internet, as encryption to internet destinations is typically handled at the application layer. Internet key exchange version 2 (IKEv2) is used for IPsec negotiation and tunnel establishment between VMware SD-WAN Edges and from VMware SD-WAN Edges to VMware SD-WAN Gateways. IKEv2 offers strong security and resiliency against attacks by bad actors.

Public Key Infrastructure (PKI)-based authentication

The VMware SD-WAN Orchestrator has a built-in certificate authority server and it manages the PKI lifecycle of the VMware SD-WAN deployment, including handling certificate signing requests (CSRs) from VMware SD-WAN Edges or VMware SD-WAN Gateways, verifying certificate authenticity and maintaining the certificate revocation list (CRL). Certificate information is exchanged via the heartbeats between the VMware SD-WAN data plane components and the VMware SD-WAN Orchestrator.

The VMware hosted SD-WAN Orchestrator has a built-in certificate authority server. The private key for the root certificate is protected by AES-256 encryption.

The passphrase used in the encryption of root certificate private key is never stored with the VMware SD-WAN Orchestrator.

Certificates issued by the VMware SD-WAN Orchestrator are used for the authentication of:

- The management plane, between the VMware SD-WAN Orchestrator and VMware SD-WAN Edge/Gateway.
- The control and data planes for establishment of IKEv2/IPsec tunnels between VMware SD-WAN Edges and between VMware SD-WAN Edge and VMware SD-WAN Gateway.

Certificates are valid for 90 days and are renewed around 60 days into the validity period of 90 days.

Authentication Modes

When provisioning a VMware SD-WAN Edge or VMware SD-WAN Gateway in the VMware SD-WAN Orchestrator, the following three authentication modes are available:

Certificate Disabled

When **certificate disabled** is used for a VMware SD-WAN Edge, the VMware SD-WAN Orchestrator will not look for certificates when processing the messages over TLS from data plane devices, however an identification token is still required by the VMware SD-WAN Orchestrator to validate data plane device authenticity. In this mode, for communication that goes between data plane devices i.e. between VMware SD-WAN Edges and from VMware SD-WAN Edges to VMware SD-WAN Gateways, these devices generate pairwise authentication keys on-demand using specific elements that are known to both devices.

Certificate Required

When **certificate required** is used for a VMware SD-WAN Edge, the VMware SD-WAN Orchestrator mandates that certificates are used for data plane authentication between VMware SD-WAN Edges and each other or VMware SD-WAN Gateways configured with certificate-required setting. For communication from VMware SD-WAN Edge to VMware SD-WAN Orchestrator, it will also require the device to present a certificate issued by the CA server on VMware SD-WAN Orchestrator.

Certificate Optional

In an existing deployment where the data plane devices are configured with **certificate disabled** and the network engineer is looking to transition to certificate-based authentication, the network engineer should select **certificate optional**, instead of **certificate required** option. **Certificate optional** will allow the VMware SD-WAN Orchestrator to accept communication from the devices without a certificate and force the device to generate a CSR for certificate enrolment. Once the VMware SD-WAN Orchestrator receives the CSR, it then issues a certificate to the device. This certificate is subsequently used for validating the authenticity of the data plane device communication with the VMware SD-WAN Orchestrator.

> **TECH TIP**
>
> Certificate Optional is the recommended setting as it allows flexibility to accommodate both pre-shared key and PKI-based authentication.

Encryption between SD-WAN components

Once authenticated with one of the three authentication modes described above, Diffie-Hellman key exchange is used to generate symmetric keys between the SD-WAN data plane endpoints, namely the VMware SD-WAN Edge and VMware SD-WAN Gateway. This key is then used to start encrypting the data using the supported algorithms ensuring the traffic going through the tunnel is secured.

Traffic between VMware SD-WAN Edges, and from VMware SD-WAN Edge and VMware SD-WAN Gateway, is secured with a IPsec-over-VCMP tunnel. IPsec uses AES-128 or AES-256 bit keys for confidentiality and up to SHA-256 for data integrity protection. Encrypted traffic is encapsulated in VCMP and UDP using source and destination port UDP 2426.

The table below summarizes the encryption options configurable in the orchestrator.

	Data plane	Management plane
Encryption	AES-CBC, Key Sizes (128, 256) AES-GCM, Key sizes (128, 256)	AES_128_GCM
Hashing Algorithms (HMAC)	SHA1, SHA-256	P-256
Key Generation/Exchange	Diffie-Hellman Groups 2, 5, 14	ECDHE_RSA

Key Takeaways:

- Communications between the data plane and the control plane are IPsec encrypted
- Communications between the data plane and the management plane are TLS encrypted
- The VMware SD-WAN Orchestrator acts as a central certificate authority (CA) server

Security of VMWare SD-WAN Solution Components

VMware SD-WAN Edge

VMware's SD-WAN Edge solution is built on top of a custom Linux operating system distribution with an up-to-date, hardened kernel. All non-essential services, utilities, and accounts have been removed. If a service or utility is necessary, security configuration best practices have been implemented to reduce the attack vector.

The VMware SD-WAN Edge implements control plane firewalling and connection rate limiting to harden it against man-in-the-middle and denial of service attacks. VMware SD-WAN Edges should be configured not to respond to connections on local ports (see Chapter 6: *Securing SD-WAN Deployment* for more detail).

A VMware SD-WAN Edge configured for dynamic branch-to-branch VPN (configured in the VMware SD-WAN Orchestrator under `Configure > Profiles > Cloud VPN` and applied to Edges that are a member of that profile) will listen on UDP port 2426 for inbound tunnel connections from other Branch Edges with configuration. While the use of this open UDP port is limited to only VMware SD-WAN Edges that are authorized to connect with the appropriate IKEv2 credentials, some highly secure organizations may still choose to disable this feature, thereby leaving no ports open on the VMware SD-WAN Edge.

> **TECH TIP**
>
> If you require branch VMware SD-WAN Edges (those not in use as a Hub) not responding to inbound connection attempts, then avoid using **Dynamic Branch to Branch VPN** in the profile that applies to such Edges. Branch-to-branch traffic can still be enabled, but branch-to-branch traffic will be via the configured hub or gateway.

VMware SD-WAN Hosting

The VMware cloud-hosted SD-WAN service provides VMware SD-WAN Orchestrators and VMware SD-WAN Gateways running in secure SSAE Type II data centers and Tier 1 cloud data centers. Network ports are locked down to the minimum required, which for the VMware SD-WAN Gateway, only UDP 2426 traffic is permitted; and for the VCO, only TLS1.2 (TCP/443) traffic is permitted. All other ports are blocked and all traffic to these ports is silently discarded.

As part of a comprehensive threat management approach, VMware conducts various types of security penetration testing and vulnerability scanning including internal and third-party engagement. Such testing is carried out periodically and whenever the underlying product architecture undergoes significant changes. Vulnerabilities identified are assessed and patched based on the criticality of the scan result. There are three categories of scan results: Confirmed Vulnerability, Potential Vulnerability, and Information Gathered. Service impacting vulnerabilities are fixed and patched within 24 hours of identifying the issue.

> Key Takeaway:
>
> - The VMware SD-WAN Orchestrators and VMware SD-WAN Gateways which comprise the VMware hosted SD-WAN service are hardened and continuously monitored for threats and vulnerabilities.

Inspection of User Traffic

Multiple options exist within the VMware SD-WAN solution for the inspection of user traffic as it passes through the VMware SD-WAN Edge appliance. Inspection for the purpose of threat detection and prevention (IDS/IPS, anti-malware, URL filtering) is available either **locally** in the VMware SD-WAN Edge via service chaining through a firewall virtual network function; or **remotely** by steering user traffic to a cloud-hosted security service by policy

The following describes the local and remote user traffic inspection options in more detail.

Local Inspection via Virtual Network Function

Organizations with a security requirement to deploy a firewall locally at the branch and at the same time consolidate the branch hardware footprint, have the option to leverage a firewall **virtual network function** or VNF directly on specific models in the VMWare SD-WAN Edge device range.

The VMware SD-WAN integrates with a supported list of firewall VNF vendors by having the firewall running as a virtual machine on the VMware SD-WAN Edge. Using a VMware SD-WAN Edge model that is VNF-ready, the VMware SD-WAN Orchestrator can be used to drive the initial VNF configuration, service chaining of the VNF inline with user traffic, and in addition, licensing the firewall VNF via API calls to the firewall VNF provider's licensing server. The VMware SD-WAN Orchestrator also provides ongoing monitoring of the firewall VM status and resource utilization.

Note, however, that firewall licensing, policy configuration and software lifecycle are managed by the firewall vendor's central management server. Figure 6.2 illustrates this integration.

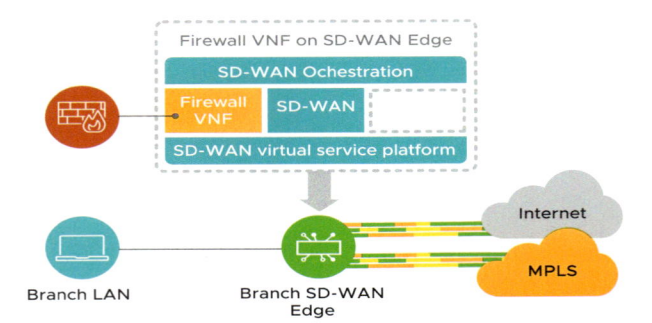

Figure 6.2: VMware SD-WAN Edge with a service-chained firewall virtual network function

The high-level workflow of bringing up the firewall VNF in the VMware SD-WAN Edge is a three-step process:

1. Add a **license server** name, **API key** and **Auth Code**, then validate the code in the **VNF License Configuration** dialog under `Configure > Network Services`

2. Deploy the Edge VNF in the **Edge VNF Configuration** dialog under `Configure > Network Services`

3. Complete any additional steps required on the firewall VNF management server such as authorising the new VNF, defining and publishing an appropriate security policy.

Traffic flow when firewall VNF integrates in the SD-WAN Edge

Figure 6.3 shows traffic flow from the branch LAN client machine

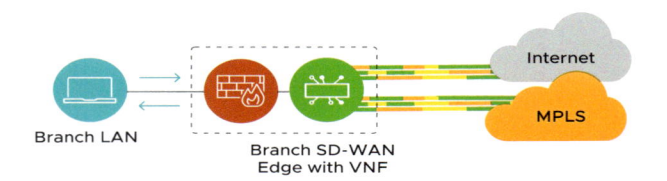

Figure 6.3: Traffic flow with VMware SD-WAN Edge has firewall VNF deployed

The insertion of the firewall VNF is always a Layer-2 insertion, such that the firewall VNF is sitting between the client LAN subnet and the LAN side of the SD-WAN Edge service. When a client machine initiates a flow to the internet, the following steps occur:

1 The client machine request arrives at the VMware SD-WAN Edge LAN interface.

2 The firewall VNF inspects the traffic flow and applies the appropriate threat prevention/detection policy as configured in the firewall VNF solution.

3 If the traffic flow is permitted, packets will arrive at the LAN side of the VMware SD-WAN Edge service and be sent to the appropriate destination based on the VMware SD-WAN routing table and **business policy** and **link steered** as required.

4 Return traffic, heading back to the client machine, is passed from the VMware SD-WAN Edge service to the firewall VNF for inspection as per the firewall VNF security policy.

In this way, the firewall VNF secures branch user traffic and mitigates threats from malicious sources. The firewall VNF has visibility into traffic to and from users in the branch before SD-WAN encryption is applied and so is able to perform meaningful inspection. The VMware SD-WAN Edge then performs all of its normal functions such as encryption, per-packet link steering, SaaS acceleration, QoS and packet loss remediation.

Segmentation-aware Integration

The VMware SD-WAN solution implements segmentation such that individual LAN side interfaces or VLANs can be placed into one or more isolated routing tables on the VMware SD-WAN Edge and the isolation is maintained over the WAN. Certain firewall VMware SD-WAN integrated VNF solutions provide a virtual wire per segment. For the segment in which the firewall VNF is inserted, there is a unique VLAN ID assigned and the firewall VNF policy is defined using this VLAN ID. Traffic from the VLAN which interfaces within these segments is tagged with the VLAN ID allocated for that particular segment.

Since every virtual wire is independent and isolated, overlapping IP addresses can be used in each virtual wire. This makes the firewall VNF fully compatible with the VMware SD-WAN segmentation. Segmentation is covered in detail in its own chapter.

Figure 6.4 shows firewall VNF inserted into Segment 1 using VLAN 200 and Segment 2 using VLAN 300. The Firewall VNF then passes traffic to the VMware SD-WAN service with appropriate segmentation tags via an additional interface (not shown).

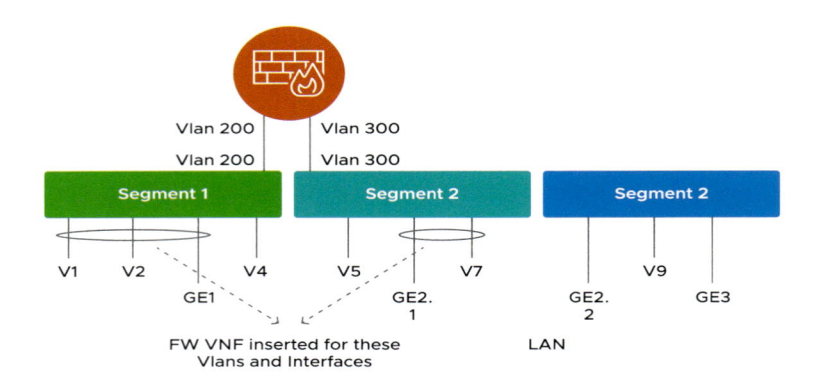

Figure 6.4: Segmentation-aware firewall VNF

Remote Inspection via a Cloud-Hosted Security Service

In a traditional WAN deployment, access to internet and SaaS applications is typically provided via internet circuits at the data center and a default route advertised into the WAN backbone from the data center to attract internet-bound traffic to the data center. While VMware SD-WAN can achieve the same flow via advertisement of a default route into the overlay, it is highly recommended to leverage business policy-based redirection to achieve better scale and application performance.

A quick recap of the order of operation as packets traverse through VMware SD-WAN Edge is depicted in Figure 6.5.

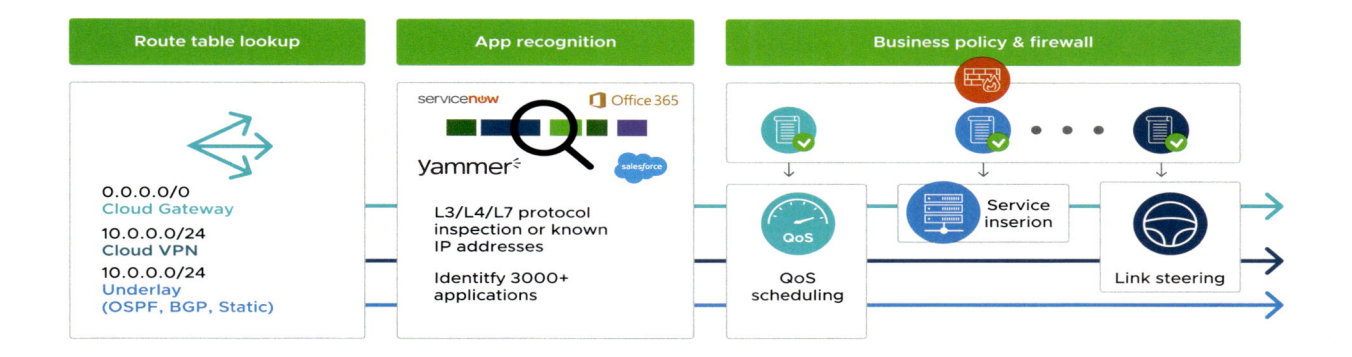

Figure 6.5: VMware SD-WAN Edge order of packet processing

The important concept to understand here is that a route lookup is performed before a business policy lookup. In order to take advantage of a policy-based re-direction option provided by business policy, it is required that default route either not be advertised from the data center LAN towards the VMware SD-WAN Hub Edges, nor block the default route inbound at the VMware SD-WAN Hub Edge.

> **TECH TIP**
>
> In the VMware SD-WAN Orchestrator go to `Test & Troubleshoot > Remote Diagnostics > Route Table Dump` and verify that you only have cloud-type routes for the 0.0.0.0 default and DO NOT have any Edge type routes for the 0.0.0.0 default in the VMware SD-WAN Edge local routing table. This example shows only cloud-type routes with the first one coming from cloud gateway (VMware hosted SD-WAN Gateway) and thus Business Policy will be applied for traffic destined to the internet.

| 0.0.0.0 | Global | 0.0.0.0 | Cloud | 0 | TRUE | Cloud Gateway |
| 0.0.0.0 | Global | 0.0.0.0 | Cloud | 5 | TRUE | GE3 |

Traffic between SD-WAN sites is sent via the SD-WAN Overlay using the best available encrypted path between those sites by default. For internet-destined traffic, the network architect can use Business Policy found in `Configure >`

Security 135

`Profiles|Edges > {profile_name}|{edge_name} > Business Policy` to decide on a per-application, per VLAN, per port or per source/destination network basis to either:

1. Backhaul traffic to specific VMware SD-WAN Edge **Backhaul Hubs** which in turn will forward the traffic to their next hop for internet breakout and perimeter security. This is done in an VMware SD-WAN Orchestrator business policy rule by selecting `Internet Backhaul > Backhaul Hubs > {specific hub(s)}` - see Figure 6.6. Note that `Internet Backhaul` will only be selectable if in the `Match` portion of the rule, under `Destination`, the `Internet radio button` is selected (not shown).

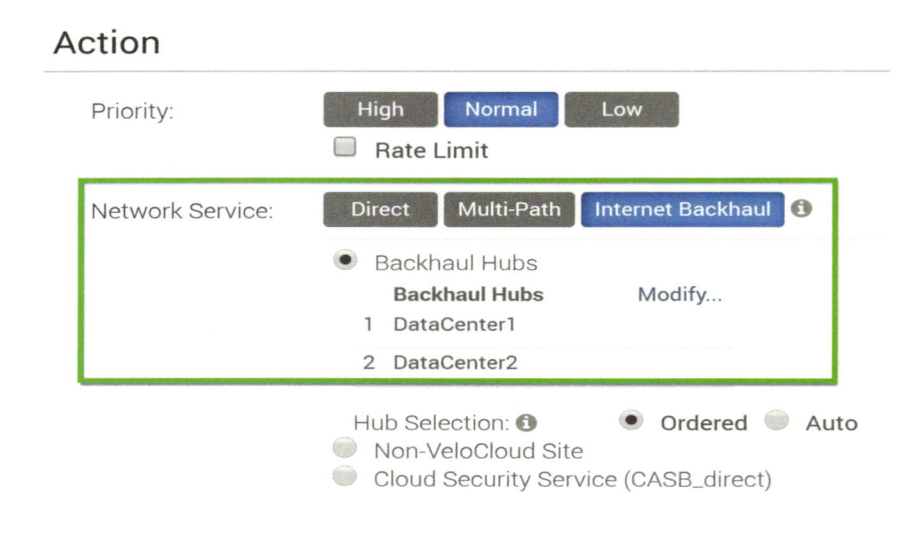

Figure 6.6: Configuring Internet Backhaul Hubs in a business policy rule

2. Backhaul to a pre-defined **Cloud Access Security Service/CASB** via a VMware SD-WAN cloud-hosted Gateway. With this method, the VMware SD-WAN Edge makes use of dynamic multi-path optimization to the cloud-hosted Gateway to mitigate congestion and latency issues, often seen over the internet. This is configured in the VMware SD-WAN Orchestrator business policy rule by selecting `Internet Backhaul > Non-VeloCloud Site > {pre-configured security service}` as shown in Figure 6.7.

Figure 6.7: Configuring internet backhaul to a CASB service called "CASB_via_Gateway" in a business policy rule

3. Breakout directly to the internet from the VMware SD-WAN Edge by selecting `Direct` option as shown in Figure 6.8. As with other traffic steering methods discussed in this chapter, this method only applies to the VLANs, IP addresses, ports or applications selected in the `Match` portion of the same rule. Application access can additionally be permitted or denied using the VMware SD-WAN Edge native firewall which can match on individual applications using L7 deep packet inspection and over 3,000 application signatures.

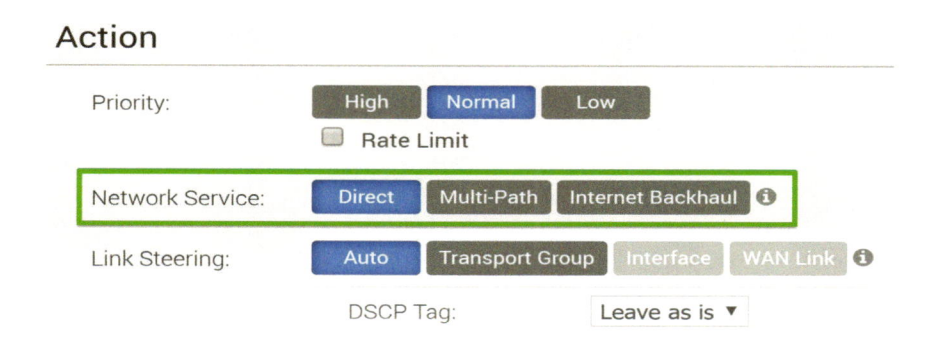

Figure 6.8: Configuring direct internet breakout in a business policy rule

Security 137

4. Redirect to the internet via the VMware SD-WAN Gateway by selecting the `Multi-Path` option as shown in Figure 6.9. For trusted SaaS destinations, internet traffic can receive DMPO benefits between VMware SD-WAN Edge and the VMware SD-WAN Gateway. Traffic is then port address translated (PAT) via the VMware SD-WAN Gateway's public IP address.

Figure 6.9: Configuring multi-path internet breakout in a business policy rule

5. Redirect traffic to a pre-defined **Cloud Access Security Service/Broker** or **CASB** directly from the VMware SD-WAN Edge via an IPSec tunnel. This is configured in an VMware SD-WAN Orchestrator business policy rule by selecting `Internet Backhaul > Cloud Security Service {name_of_CASB_service}`. Figure 6.10 shows an example selecting the **Cloud Security Service** called `CSB_Direct` already defined in the VMware SD-WAN Orchestrator under `Configure > Network Services` (not shown).

Action

Priority: High | **Normal** | Low
☐ Rate Limit

Network Service: Direct | Multi-Path | **Internet Backhaul** ⓘ
○ Backhaul Hubs
○ Non-VeloCloud Site
● **Cloud Security Service (CASB_direct)**

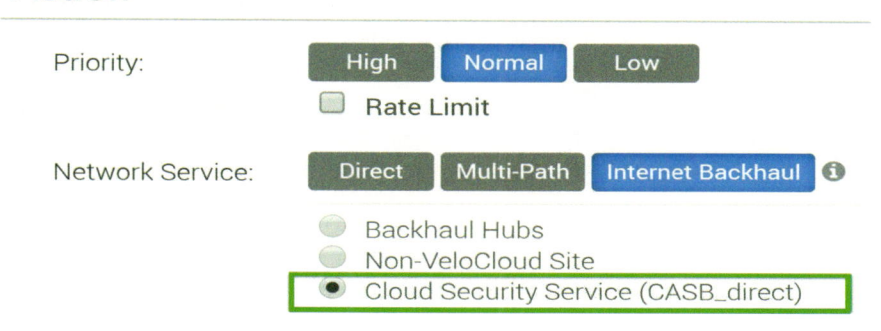

Figure 6.10: Configuring a direct IPSec tunnel directly from the VMware SD-WAN Edge to a Cloud Security Service called "CASB_direct"

With the major shift to SaaS-based applications, it is highly likely that a combination of the internet breakout methods discussed in this chapter is required. With VMware SD-WAN Edge business policy, the network engineer can easily, and securely, direct traffic types bound for the internet, along the optimum path for further best security, best performance, or a mix of both.

The following example in Figure 6.11 is a real-world scenario where a network architect needed to cover the following requirements:

1. Highly critical business application traffic needs to be backhauled to a VMware SD-WAN Edge in the data center where it can be forwarded through a corporate firewall for compliancy reasons before going to the internet.

2. Office 365 traffic should benefit from VMware SD-WAN built-in optimization over multiple internet links and sent via a VMware SD-WAN Gateway.

3. Business internet apps should be redirected directly from the Edge over an IPSec tunnel to a nearby Cloud Security Service.

4. Guest Wi-Fi users at this branch can break-out directly to the internet.

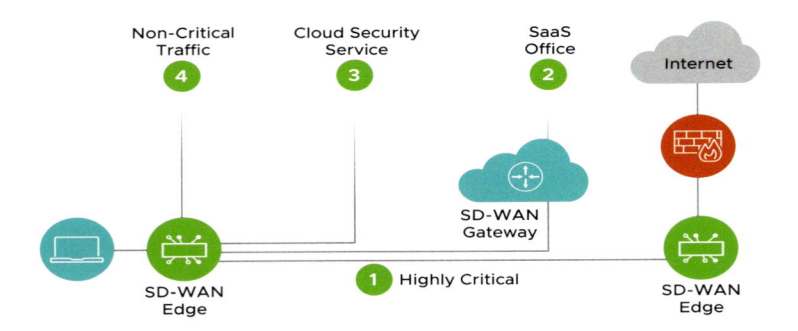

Figure 6.11: Application access via VMware SD-WAN Edge is directed to the appropriate path by policy

The path selection by application discussed above can be realized by a combination of business policy rules shown in Figure 6.12. Rule no. 1 covers requirement one; Rule no. 2 covers requirement two and so on.

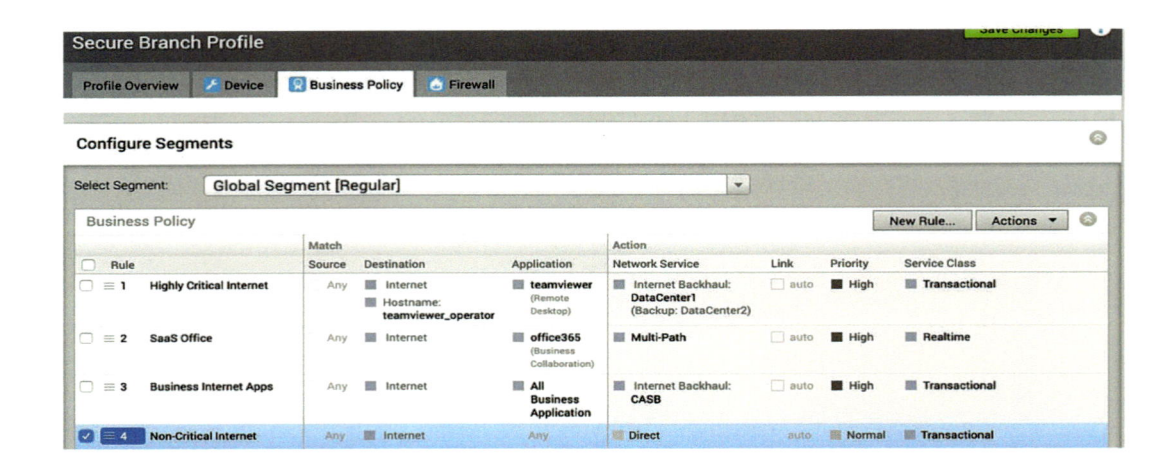

Figure 6.12: Business policy rules used to steer application traffic to a desired path

T E C H T I P

Business policy rules are processed and executed sequentially, so the order of rules is important.

Key Takeaway: VMware SD-WAN gives you the flexibility to inspect user traffic via various means that suit your budget, security requirements, and desired application performance outcomes.

VMware SD-WAN Segmentation

The Need for Segmentation

Network segmentation allows an organization to compartmentalize application traffic in the network infrastructure, separating users and assets belonging to different security zones. Some of the common use cases for segmentation include separating guest traffic from corporate traffic, and isolating credit card transactions from all other traffic types.

In traditional networks, segmentation requires complex CLI-based configuration or dedicated hardware. More importantly, administration of the VLANs and VRFs increase the workload of the network engineer significantly. The biggest drawback of traditional network segmentation techniques is they are locally significant to the device, and the network engineer has to provision every network device in the path, hop-by-hop, to extend the segmentation end-to-end. This results in network engineers having to spend significant effort to ensure the segmentation is implemented correctly and policy compliance is being met.

Below are a few common use cases for network segmentation:

- Line-of-business separation by departments for security and audit.
- User data separation: guest traffic, Payment Card Industry (PCI) data, employee traffic.
- Mergers and acquisitions: network segmentation is used to allow overlapping IP addresses and secure access to shared assets.
- An external party such as a partner needing access to a subset of information in the corporate data.

VMware SD-WAN provides edge-to-edge segmentation that can be centrally provisioned and pushed to some or all SD-WAN branch sites, making it scalable, easy to manage, and cost-effective. Each segment is treated as a separate configuration entity having its own set of cloud VPN, business policy, firewall, and QoS configuration elements.

How VMware SD-WAN Segmentation Works

Let's begin with a diagram to explain how segmentation works. Figure 6.13 depicts a VMware SD-WAN Edge in the data center acting as a hub, with three network segments behind the hub - **Voice, Guest, and PCI**. There are three branch sites, and each has a different combination of the segments.

- Branch 1 has Voice and PCI segments
- Branch 2 has Voice and Guest segments
- Branch 3 has Voice, Guest, and PCI segments

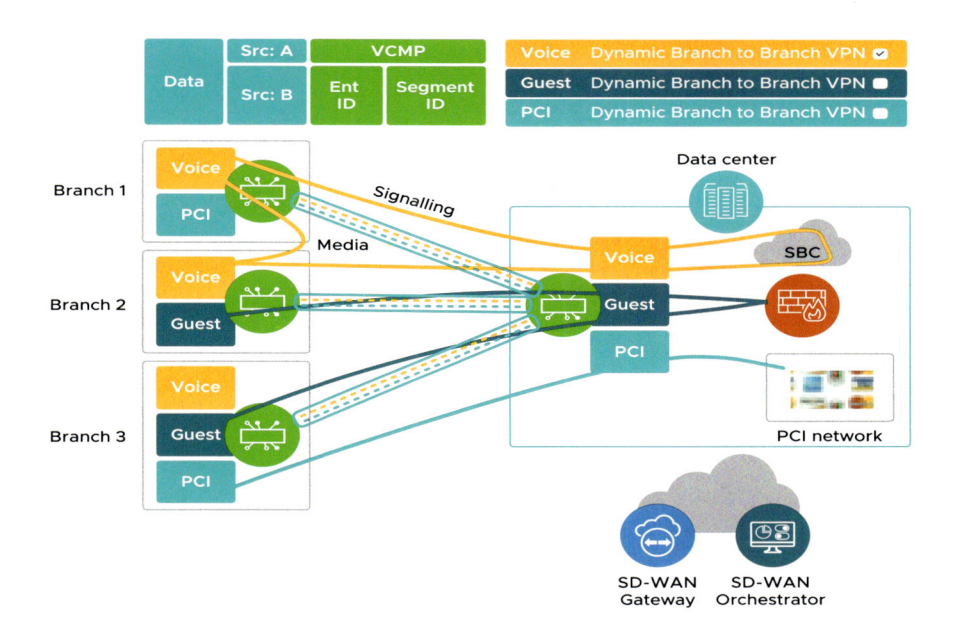

Figure 6.13: VMware SD-WAN segmentation

The three segments - **Voice, Guest,** and **PCI** are isolated using both VLANs on the LAN side and VMware SD-WAN segmentation on the WAN side of the VMware SD-WAN Edge. With this configuration, a network endpoint connected to one segment is unable to communicate with another network endpoint on another segment. Of course, a device external to the SD-WAN solution such as a firewall or router provides inter-segment connectivity.

The number of VCMP tunnels from a branch to the Hub is independent of the number of segments that are configured in the VMware SD-WAN solution. The header for the VCMP tunnel features a segment ID field which allows each segment to be uniquely tagged for separation, allowing the segments to share a common VCMP tunnel. This design offers an elegant mechanism for extending network segmentation edge-to-edge without compromising scalability of the overall SD-WAN system.

Each segment can have a different cloud VPN topology. In this scenario, guest and PCI segments have hub-and-spoke topology, while the voice segment has the dynamic branch-to-branch VCMP tunnel, enabling a lower latency path which results in better voice quality.

Configuration of VMware SD-WAN Segmentation

The configuration of the segmentation is simple as it only requires three easy steps.

First, create one or more new `Segments` as shown in Figure 6.14

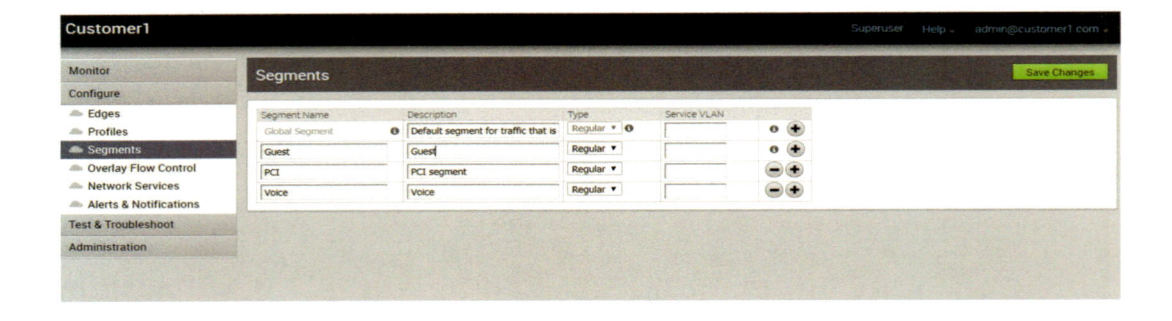

Figure 6.14: Create segments

Second, edit the profile containing sites that need segmentation e.g. branch sites and select the new segment(s) that should be available within the profile as shown in Figure 6.15. Click `Configure > Profiles > {Profile Name} > Device > Select Profile Segments > Change`

144   SD-WAN 1:1

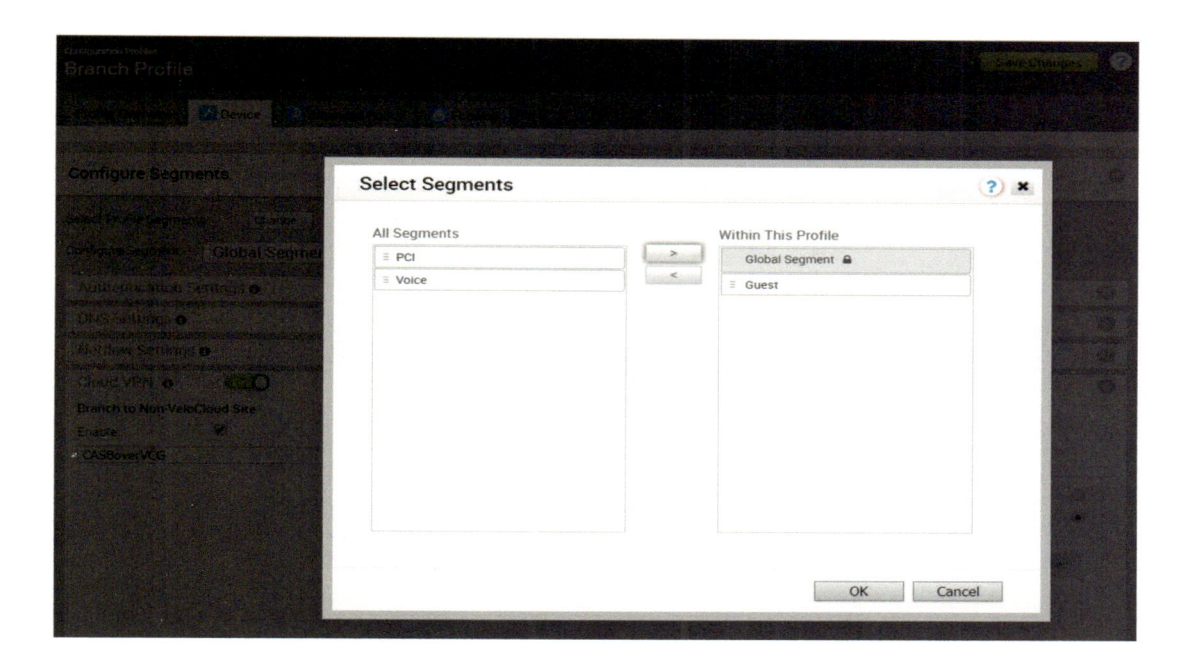

Figure 6.15: Select which segment is available to a profile

Finally, for each VMware SD-WAN Edge model type listed in the VMware SD-WAN Orchestrator profile, define the VLAN(s) that belong to the new segment and map the VLAN to the segment by selecting it in the `Segment` drop down as shown in Figure 6.16.

> **TECH TIP**
>
> Routing in the VMware SD-WAN Edge is possible between VLANs within the same segment, but not between VLANs in different segments.

VLAN

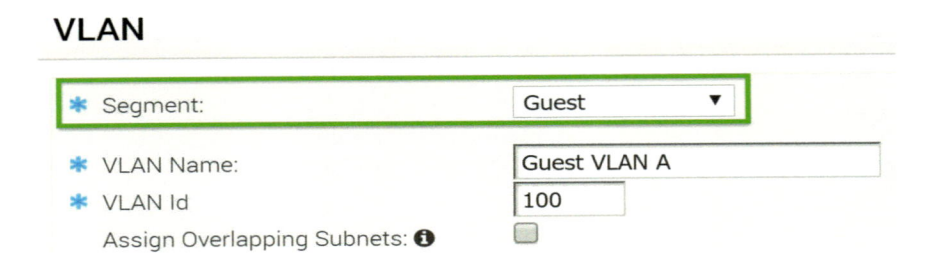

✱ Segment:	Guest ▼
✱ VLAN Name:	Guest VLAN A
✱ VLAN Id	100
Assign Overlapping Subnets: ⓘ	☐

Figure 6.16: Create a new VLAN and assign it to the appropriate segment

The new VLAN and mapping to the segment is then pushed to all VMware SD-WAN Edges that are a member of the profile.

Similar configuration can be done on a routed LAN port (WAN Overlay disabled) as shown in Figure 6.17.

Interface: GE3 ☑ Override Interface
- Interface Enabled: ☑
- Capability: Routed ▼
- Segments: Guest ▼
- Addressing Type: DHCP ▼
  - IP Address: n.a
  - CIDR prefix: n.a
  - Gateway: n.a
- WAN Overlay: ☐

Figure 6.17: Assigning the appropriate segment to the desired routed interface.

Additional settings such as business policy and firewall rules are available on a per-segment basis in the profile by selecting the segment you wish to configure in the drop down menu at the top of the profile. E.g. `Business Policy > Select Segment > {drop down menu}`

146   SD-WAN 1:1

> Key Takeaway: VMware SD-WAN offers an elegant mechanism for extending network segmentation edge-to-edge without compromising scalability of the overall SD-WAN solution. At the same time, each segment can be configured independently for VPN topology, business policy, and firewall rules.

Securing SD-WAN Deployment

VMware SD-WAN Required Ports and Protocols

The VMware SD-WAN solution makes use of different protocols and ports depending upon the connection type ie. management, control or data plane connection, as outlined in earlier chapters. The Figure below details the connectivity between the VMware SD-WAN Edge, VMware SD-WAN Gateway, and VMware SD-WAN Orchestrator.

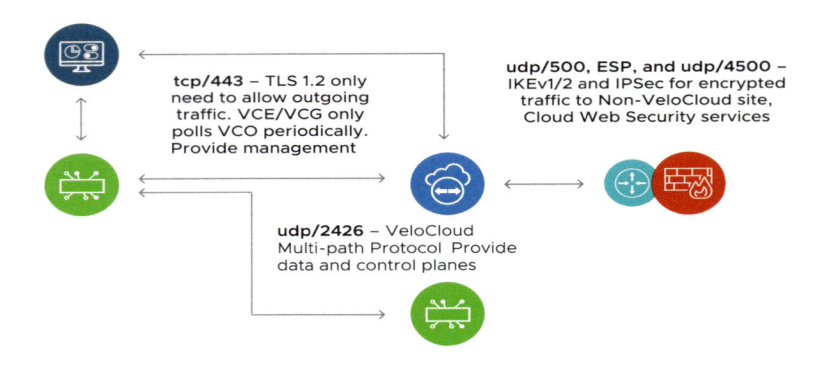

Figure 6.18: Protocols and ports in use for communication between SD-WAN components.

In a data center deployment, the following ports should be allowed on the firewall from the SD-VMware WAN Edge appliance(s):

- TCP 443 (outbound)
- UDP 2426 (inbound & outbound)
- DNS (53) and NTP (123) (outbound)

In a branch deployment, the following ports should be allowed in case a firewall is placed in front of the VMware SD-WAN Edge:

- TCP 443 (outbound)
- UDP 2426 (outbound)
- DNS (53) and NTP (123) (outbound)

In the event the above referenced ports are not opened on the firewall, in particular inbound UDP 2426 on the data center firewall, the branch edges will not be able to establish tunnels to the data center edge (Hub). This can be verified by going to `Test & Troubleshoot > Remote Diagnostics > {Edge name} > List Paths > Peer > {drop down menu will not show Hub edge as peer}`

List Paths
View the list of active paths between local WAN links and each peer.

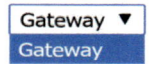

Peer   Gateway ▼
        Gateway

Figure 6.19: There is no VMware SD-WAN Edge available under List Paths which indicates there is no branch to hub tunnel

Roles-based Access Control Architecture (RBAC)

The VMware SD-WAN Orchestrator is designed for multi-tenant service providers and enterprise environments alike. It implements a Roles-based Access Control (RBAC) method to provide a secure multi-tenancy platform that ensures authenticated users see, and do, only the functions allowed by their assigned role.

The VMware SD-WAN Orchestrator supports three organizational tiers, each with its own subset of roles and privileges to properly segment users based on their responsibility. The three organizational tiers as shown in Figure 6.20 are:

- Operators of the solution have full access to all accounts provisioned in the system as well as system management of the VMware SD-WAN

Orchestrator and VMware SD-WAN Gateways management and assignment.

- Partner accounts are allowed to manage only the enterprise accounts that were created or assigned to the partner. Optionally, partner accounts can manage on-premises VMware SD-WAN Gateways but the partner does not have access to VMware SD-WAN Orchestrator system functions.

- Customer (also referred to as Enterprise) accounts provide exclusive access to all the edges and policies of a single enterprise and are typically provided to end-users of the SD-WAN solution.

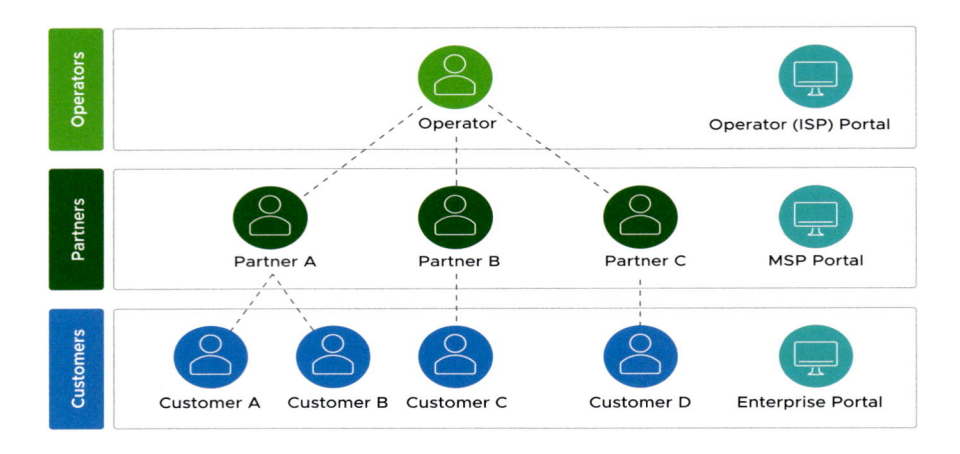

Figure 6.20: VMware SD-WAN Orchestrator organizational tiers.

A user can be assigned to one of these three tiers. A user type defines the scope of access within a given organizational tier. The user types at the customer level are:

- Superuser: User can view and create additional admins.
- Standard Admin: User can view and manage their Customer.
- Customer Support: User can view (but not manage) their company's network.
- Enterprise Read Only: Enterprise Read Only User

The VMware SD-WAN Orchestrator utilizes a native database to authenticate accounts but can also use an external RADIUS server, SSO or Two-Factor authentication for this validation. The VMware SD-WAN Orchestrator supports the following authentication methods for both hosted and on-prem deployments.

1. **Two-Factor authentication**: available for operator, partner and enterprise users. When two-factor authentication is enabled, user login attempts are verified with a PIN sent via a text message to a user's mobile phone. The user provides the PIN back to the VCO where it is validated to match a stored value for successful authentication

2. **RADIUS-based authentication**: available for operators. In the case of RADIUS, the user role mapping is defined via the RADIUS server and passed back as attributes during VMware SD-WAN Orchestrator authentication.

3. **Single-Sign On (SSO)**: available for operator, partners and enterprise users. The VMware SD-WAN Orchestrator supports standards-based Open ID Connect (OIC) Relay Party mode. When a user logs in at the enterprise level, an option to login to ID provider of choice, e.g. OneLogin, Okta, is provided. When the user chose to sign-in to the ID Provider of choice, the VMware SD-WAN Orchestrator portal automatically redirects the user to the SSO provider login portal. When the user successfully authenticates with the ID provider, an OIC ID-Token will be granted which provides the access to VMware SD-WAN Enterprise portal, along with correct customer ID and URL for automatic redirect and instant login. Figure 6.21 below shows the process to login using the `Single-Sign On` feature.

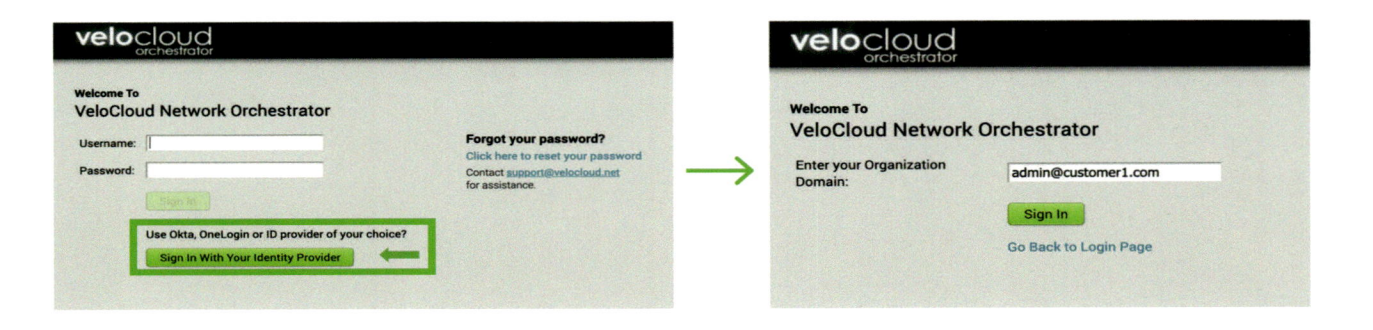

Figure 6.21: SSO login utilization on the orchestrator.

Edge Security Best Practices

From a security standpoint, following are the best practices to employ in securing the VMware SD-WAN Edge.

1. Ensure `Support Access` and `Local Web UI Access` are set to `Deny All`. This can be set at the profile level and pushed down to all the VMware SD-WAN Edges in the enterprise. If Local Web UI access is needed, then provision it at the VMware SD-WAN Edge level and allow only specific IPs to access the UI. The same goes for `SNMP access`. `Support, SNMP and Local Web UI` accesses are all set to `Deny All` by default.

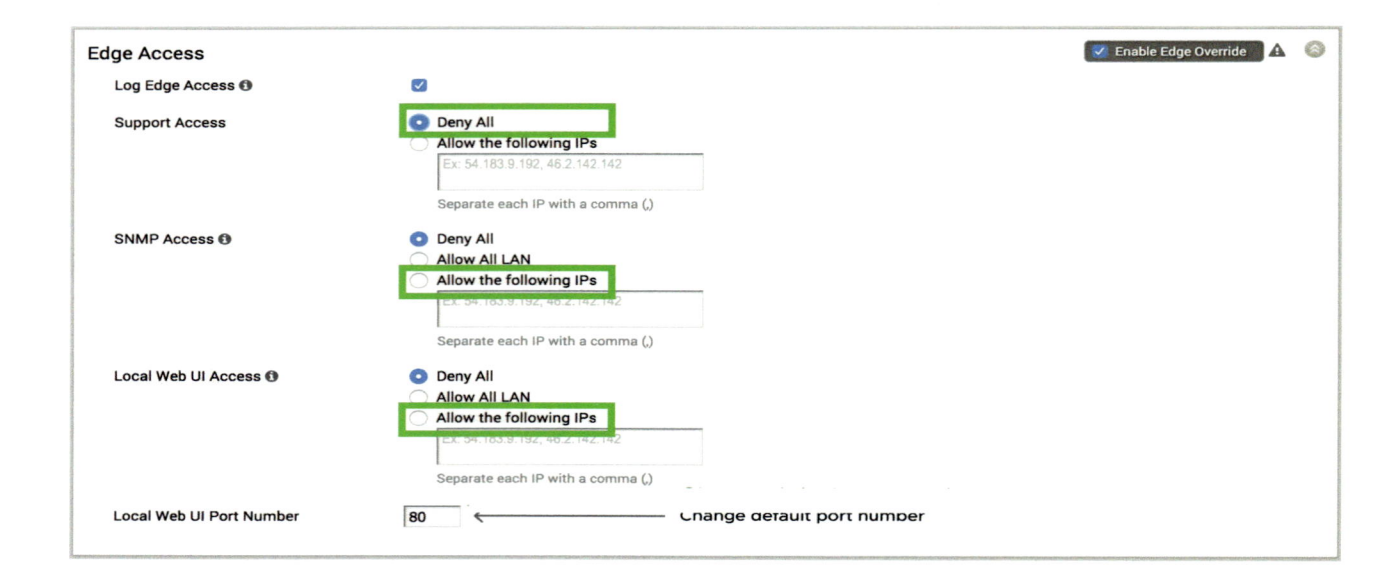

Figure 6.22: Recommended Edge Access settings—Deny All where possible

2. In conjunction with the security settings for `Edge Access`, change the username/password from the default admin/admin to something more secure such as a 12 character password with sufficient complexity. This can be done in `Configure > Profiles > {profile_for_edges} > Profile Overview tab > Local Credentials`. Figure 6.23 shows the `Local Configuration` Credentials dialog. Note that local credentials are only needed if you have allowed access to the `Local Web UI` in step 1. above.

Local Configuration Credentials	✖
* User — admin	
* Password — •••••••	
Submit Close	

Figure 6.23: Dialogue to change the Local UI login credentials.

> Key Takeaway: The VMware SD-WAN Orchestrator is the one-stop-shop to manage VMware SD-WAN Edge configuration. Use of the VMware SD-WAN Edge local services - apart from SNMP - are not recommended for in-production devices and should only be used for troubleshooting specific issues under the guidance of VMware SD-WAN support.

Rodney leaves goat yoga feeling refreshed and armed with some knowledge from Yves. The organization Yves works for (how does one get a job as a musician at a tech company?) overcame similar challenges by combining cloud-based security and centralized security at the data center. Just like Allofu, it's rarely a one size fits all world. As Rodney walks into the building the next morning starting on latte number two for the day, he is greeted by Hans, frown and all.

Hans is, of course, more than a little disappointed that Rodney has not magically addressed his concerns overnight. Rodney suggests that Hans join him in the meeting with Leia and Raycliff over lunch to learn firsthand what VMware suggests.

Alvina looks up from her desk as she finishes her lunch and sees Rodney, Hans, and the VMware team exiting the conference room. Did Hans just smile? Surely not and there…it's gone. Alvina really wants to know how the meeting went and if her team has been able to address the concerns raised by Hans—she really has had enough of these nightmares…maybe this weird goat yoga thing Rodney does would help get her mind off all these issues? Then she thinks yoga maybe, but yoga with farm animals just is a bit too weird.

Alvina requests an impromptu meeting with Rodney and Hans for an update.

Rodney is very happy to report that VMware SD-WAN has multiple ways of addressing Hans' security concerns. Rodney goes over the summary of the VMware SD-WAN's security options and the potential for Allofu to employ these for different application flows.

Alvina notices Hans' frown is starting to become less visible, although he is still worried about internet access in branches handling sensitive customer data. Rodney reminds Hans that with segmentation (inherent in VMware SD-WAN), they could isolate regular corporate users from sensitive data. Hans does mention that it was "pointless" to send O365 traffic through his firewalls and the ability to by-pass his firewalls for select traffic such as this is powerful and will help him with the scale on his appliances.

Hans appears to actually smile, but it's not entirely clear that these muscles have been exercised in recent history. Hans would like to extend the pilot to include cloud security and integrating his data center firewalls with the SD-WAN pilot.

7. Migration

Rodney is cautiously optimistic with how smoothly the SD-WAN pilot sites have rolled out and with how they have performed. Rodney's team has now successfully rolled out multiple VMware SD-WAN Edge deployment topologies and triumphantly proven single link, multiple links, and even an LTE connection for Henrik. Henrik experienced a brief outage, but a squirrel chewing through the power cord is truly outside of Rodney's domain of responsibility. Rodney has now begun to shift his focus to the "when" instead of the "if" he will migrate Allofu's network to SD-WAN. Hans in his own unique way (he nodded and didn't say "No") has given his endorsement to move forward.

Alvina is feeling optimistic as well but wants to see a plan from Rodney and Hans on how they would migrate the rest of Allofu's branches to SD-WAN. While the manufacturing plants may close on major holidays, there are no holidays for Allofu's branch locations. Therefore, downtime is to be avoided at all cost.
If Alvina decides to move forward with the SD-WAN rollout she will need to see a very careful plan from Rodney and Henrik to give her the confidence that they have covered all their bases.

From Traditional WAN to SD-WAN

Overview

This chapter covers the workflow a network architect should utilize to begin a migration from a traditional WAN network towards a fully functional SD-WAN network.

It begins with a description of a traditional WAN as-is architecture and the desired to-be SD-WAN architecture. The migration steps, considerations, and challenges are outlined in the following.

Current Architecture

Figure 7.1 shows a typical topology found in an existing WAN deployment.

All sites are connected via single or multiple MPLS networks, with an optional backup path from branch to data center using IPsec VPN over internet. All internet traffic is backhauled from branch to data center for inspection at the central firewall.

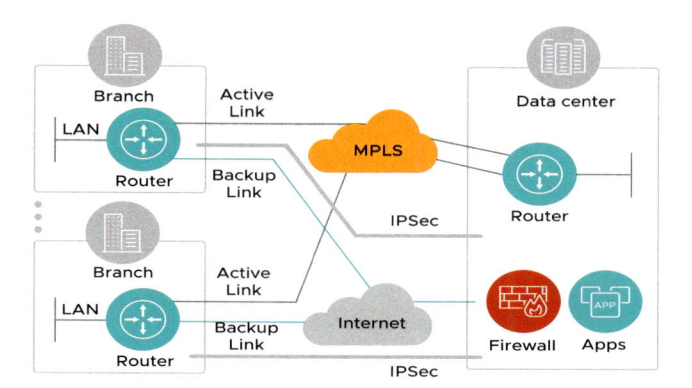

Figure 7.1: Current architecture

This design principle introduces some well-known challenges that can potentially impact end-user experience especially when deployed at a large scale.

1. A complex routing design is needed if one needs to integrate between different MPLS service providers.

2. Design lacks centralized management and monitoring.

3. Long deployment cycles for new sites due to the dependency on circuit delivery and on-site technicians required for installation.

4. Standby links are not utilized.

5. Expensive circuits such as AWS DirectConnect or Azure ExpressRoute are needed to access Infrastructure-as-a-Service (IaaS).

6. Software-as-a-Service (SaaS) user experience is poor due to backhauling of internet traffic.

7. All the above is multiplied, if the network is internationally based, with even longer lead times for international circuits, additional latency, and added complexity to maintain the network.

Target Architecture

Figure 7.2 depicts a high-level example of a successful migration to a fully deployed VMware SD-WAN environment.

VMware SD-WAN Edges and VMware SD-WAN Gateways are connected via transport-agnostic SD-WAN overlays. This architecture eliminates the dependencies on service providers in providing a consistent IP-VPN topology. Furthermore, the architecture delivers consistent reliability and quality regardless of the underlying transport, even in cases where multiple service providers are used. Applying a consistent policy to this SD-WAN Overlay from a central VMware SD-WAN Orchestrator enables the use of flexible and secure internet breakout allowing the agile use of cloud-based services. Bringing up new sites becomes as simple as connecting an internet link and powering on the devices.

Additionally, the network can be easily extended into public clouds such as AWS or Azure, utilizing virtual VMware SD-WAN Edges which are available on the respective cloud provider marketplace. This eliminates the need for expensive on-ramp connections sold by public cloud providers and enables easy consumption of their IaaS offerings.

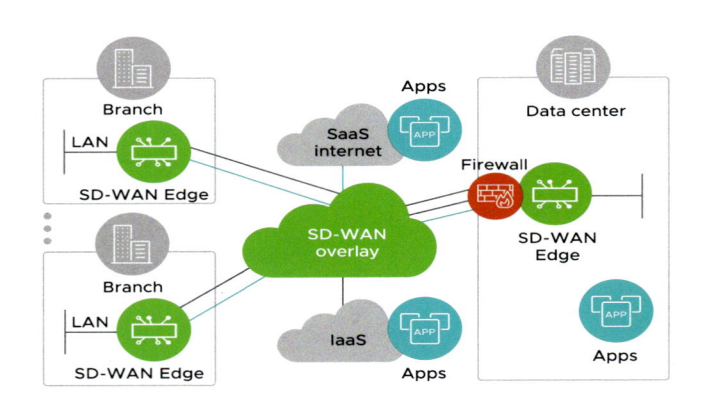

Figure 7.2: Target architecture

Routing Consideration During Migration

Business continuity is very critical for organizations during network migration, so it is very important to plan out all requirements before migration.

Following are the routing considerations to be taken into account during migration from traditional WAN to SD-WAN:

1. A way to handle different underlay types. This is to enable connectivity end-to-end.
2. A way to distinguish SD-WAN prefixes from non-SD-WAN prefixes in order to apply BGP filtering at the SD-WAN branch. This is to prevent that SD-WAN branch from becoming a transit site for non-SD-WAN prefixes, while advertising SD-WAN prefixes to non-SD-WAN sites.
3. A way to enable communication (if required) between SD-WAN and non-SD-WAN sites
4. A way to stop mutual redistribution between overlay and underlay (on branch sites). This is to prevent that SD-WAN branch from becoming a transit site for non-SD-WAN prefixes.
5. A way to redistribute overlay and underlay with a less preferred route preference in both directions (on data center/hub sites). This is to allow non-SD-WAN sites to use underlay as a primary path and overlay as a backup path to reach SD-WAN sites.

Item 1 refers to the scenario during migration where it is common to have sites with disjointed underlays. Environments that are often encountered include:

- Sites with internet only
- Sites with MPLS only
- Sites with MPLS from Provider A and sites with MPLS from Provider B.

Connectivity between sites with disjointed underlays are enabled via a transit site which is connected to both underlays. This transit site is typically a data center location which can accommodate the traffic back and forth from both underlays. As shown in Figure 7.3, traffic is sent from the VMware SD-WAN Edge to the transit site using one transport, and from the transit site to another VMware SD-WAN Edge using a different transport. Whether the traffic is sent via overlay or underlay depends on whether the respective site is already SD-WAN enabled or not.

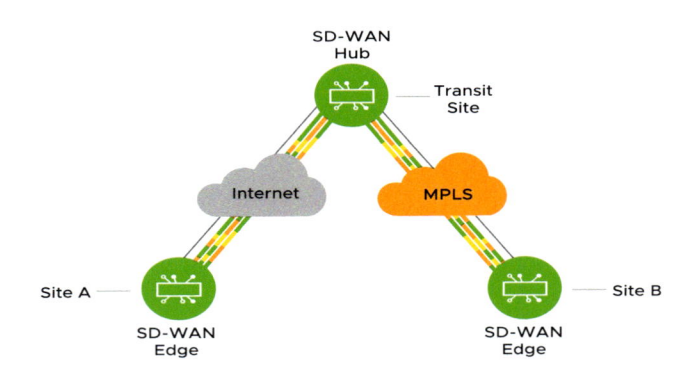

Figure 7.3: Example of disjoint underlay

Item 2 can easily be done by tagging all SD-WAN prefixes before advertising to the underlay, and subsequently filtering them out at the hybrid SD-WAN when learnt from the MPLS underlay. The tagging is done via a BGP community e.g. 12345:888. Navigate to **Configure > Profile > {Profile name} > Device tab > BGP > Edit > Add Filter**

For all SD-WAN sites, the corresponding configuration of the filter list will appear as it does in Figure 7.4:

		Filter Name	Rule Match			Rule Action	
			Type	Value	Exact Match	Type	Set
Edit | Del	1.	Allow All Inbound	Prefix	0.0.0.0/0	✗	Permit	None
Edit | Del	2.	Allow All Outbound	Prefix	0.0.0.0/0	✗	Permit	Community 12345:888 Community Additive: Enabled
Edit | Del	3.	Deny Default Routes	Prefix	0.0.0.0/0	✓	Deny	
Edit | Del	4.	Deny SD-WAN Prefixes	Community	12345:888	✓	Deny	

Figure 7.4: Recommended BGP filters and community tagging for SD-WAN prefixes

Item 3 has the following traffic flows to consider between SD-WAN and traditional MPLS sites:

An internet-only SD-WAN site will communicate with a traditional MPLS site via a transit SD-WAN Hub. Traffic is sent via the overlay from the internet-only SD-WAN site to the transit site, and via the underlay from the transit site to the traditional MPLS site. Figure 7.5 depicts this traffic flow.

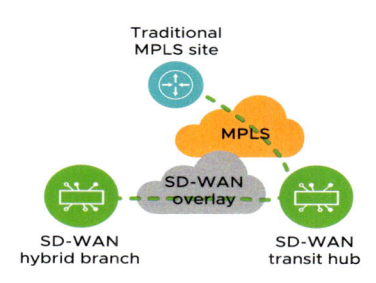

Figure 7.5: Traffic flow from internet-only site to MPLS site

To connect a hybrid SD-WAN site to traditional MPLS site, there are two options:

- Via a transit site as depicted in Figure 7.6. Note that the Figure does not represent the physical connectivity from the Hybrid SD-WAN branch to both the MPLS and the internet transports. In this case, it is recommended to turn off routing protocols on WAN links at branch sites. Figure 7.6 depicts this traffic flow.

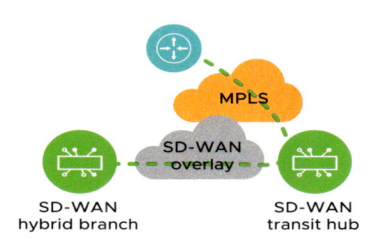

Figure 7.6: Traffic flow from Hybrid SD-WAN to MPLS site via transit hub

Migration

- Using native MPLS (underlay) between the two and using the transit site as a backup in the event the MPLS underlay is unavailable in the hybrid site. Figure 7.7 depicts this traffic flow.

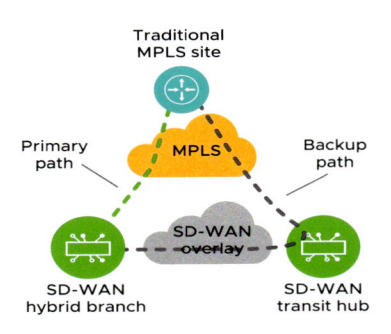

Figure 7.7: Traffic flow from Hybrid SD-WAN to MPLS site via direct underlay as primary and via transit Hub as backup

> **TECH TIP**
>
> It is best not to run routing protocols on WAN links on branch sites if there is no requirement for branch-to-branch traffic via the underlay or if connectivity between SD-WAN branch and traditional MPLS branch is via central Hub. Not running WAN side routing protocol on branch sites will greatly simplify routing as branches will only learn and advertise routes over the overlay.

Item 4 can be achieved by flagging the BGP neighbor, which is the final exit/entry point to MPLS underlay as the Uplink Neighbor for in-path insertion topology and Uplink Community for off-path insertion topology.

Please refer to Chapter 5: *Routing* for an explanation of Uplink and Uplink Community. Figure 7.8 shows an in-path SD-WAN branch where the BGP Uplink feature has been configured.

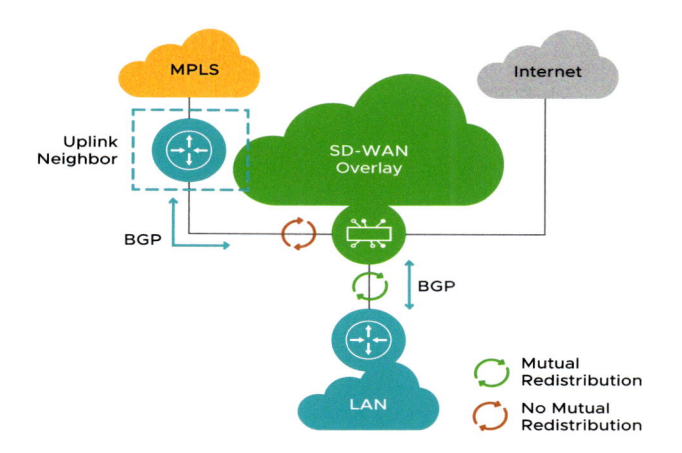

Figure 7.8: Uplink for in-path VMware SD-WAN edge insertion

Figure 7.9 shows that for an in-path VMware SD-WAN Edge, the corresponding configuration can be reached by navigating to **Configure > Edge > {Edge name} > Device tab > BGP > Edit > Additional Options > Neighbor Flag: Uplink**:

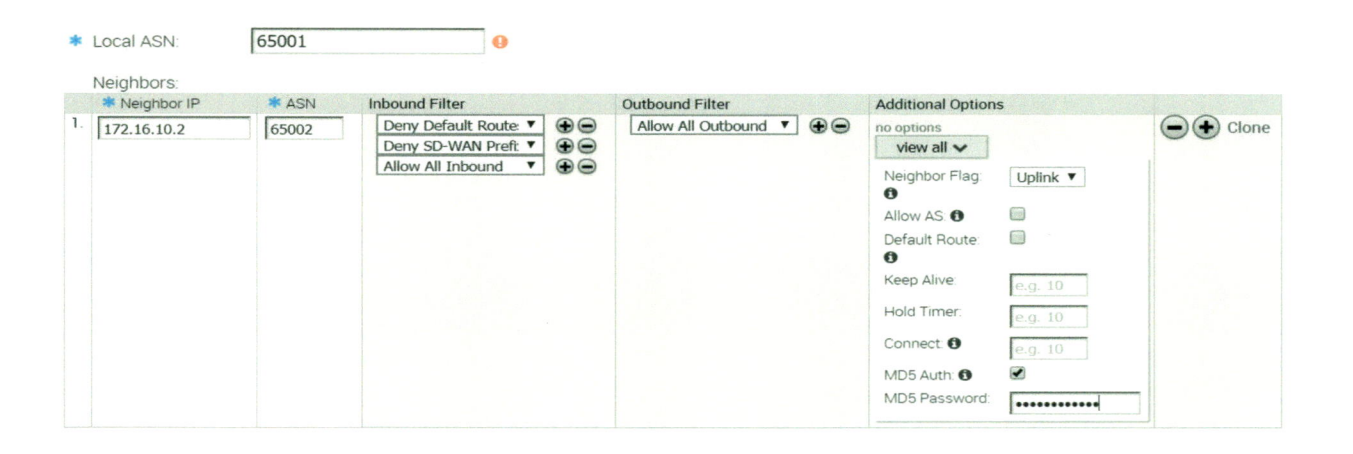

Figure 7.9: Setting the uplink flag on a BGP neighbor

Migration 163

Figure 7.10 shows an off-path SD-WAN branch where BGP Uplink Community feature has been configured.

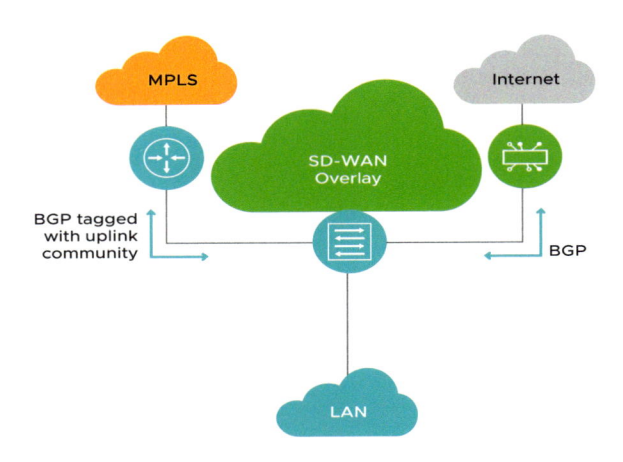

Figure 7.10: Uplink Community for off-path VMware SD-WAN Edge insertion

For the off-path VMware SD-WAN Edge, the corresponding configuration for `Uplink Community` will appear like Figure 7.11 which can be reached by navigating to `Configure > Edge > {Edge name} > Device tab > BGP > Edit > Advanced Settings > Uplink Community`:

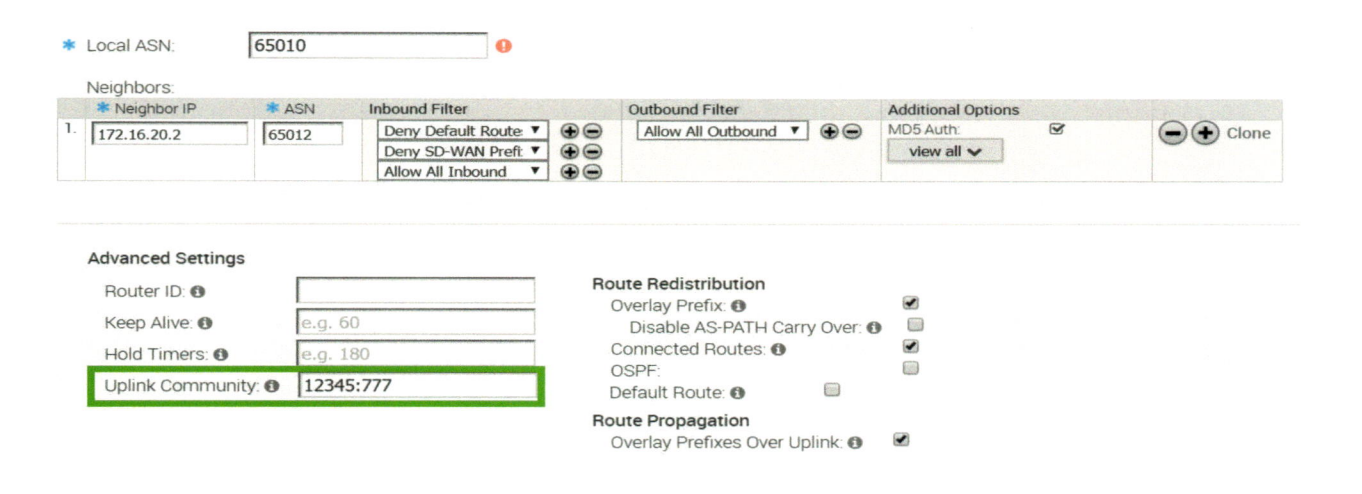

Figure 7.11: Uplink Community configuration with 12345:777 is the BGP Uplink Community and Overlay Prefixes Over Uplink is checked in the UI

Item 5 can be achieved by allowing mutual redistribution between underlay and overlay prefixes at the data center. The hybrid SD-WAN sites will learn traditional site prefixes via both the overlay (from the Hub) and via local BGP peering with the service provider PE. The default route preference setting in the VMware SD-WAN `Overlay Flow Control` will ensure that the traditional site prefixes learnt over the overlay is the less preferred routing path for traffic from SD-WAN sites to underlay prefixes advertised from traditional sites.

Routing preference from traditional sites to SD-WAN prefixes can be influenced by route manipulation technique such as BGP As-Path prepend when advertising SD-WAN prefixes from the Hub into the underlay thereby making this path less preferred when compared to the direct underlay path between traditional sites to SD-WAN branch sites. Figure 7.12 depicts the traffic flow described above.

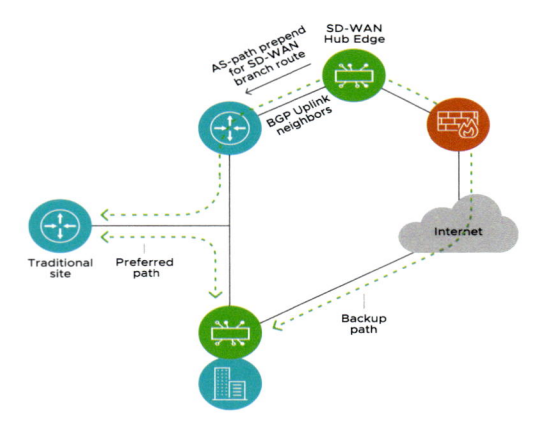

Figure 7.12: Routing consideration at the SD-WAN Hub

> **Key Takeaways:**
>
> - Enable routing on branch WAN only when absolutely needed.
> - Tag SD-WAN prefixes from all branches and data centers with a BGP community.
> - If routing is needed on branch WAN, enable the BGP uplink feature.
> - Plan traffic flows before starting the actual migration to avoid disruptions once migration has begun.
> - Decide on the traffic flow approach between SD-WAN and non-SD-WAN sites.

Migration Workflow

This chapter covers the migration workflow for transforming a traditional WAN network to SD-WAN. It also addresses the question of whether co-existence with the traditional network is required versus doing a full migration to SD-WAN. Migration steps are as follows:

Step 1. Infrastructure: Choose the VMware SD-WAN deployment model

- Which VMware SD-WAN Orchestrator and VMware SD-WAN Gateway deployment model will be used?
- Will the VMware SD-WAN Orchestrator be hosted?
- Will the VMware SD-WAN be a fully on-premises deployment?

Step 2. Data center: Deploy SD-WAN Edge in the data center

- Install and activate the VMware SD-WAN Edges at the data center.
- Enable routing protocol integration with existing network devices at the data center.
- Ensure prefixes from the data center are learnt by the SD-WAN edge but not advertised towards the data center devices.

Step 3. Branch: Deploy SD-WAN Edge at the branch

- Install and activate the VMware SD-WAN Edge in the branch.
- Define the traffic flow: Is routing needed at the branch towards MPLS WAN? How will branch users access the internet or other non-SD-WAN sites?

Step 1. Choose the SD-WAN Deployment Model

Determine the SD-WAN deployment model as outlined in Chapter 2: *Placement of Components*. This decision must be made before a network architect can begin site migration planning. The two most common considerations should be whether the deployment model is a cloud-hosted solution or a fully on-premises solution.

Step 2. Deploy Mware SD-WAN Edge in the Data Center

Upon choosing the SD-WAN deployment model, the most logical place to start with the SD-WAN deployment is the data center. This is because it has the least impact on the organization's infrastructure and traffic flow. Traffic flow will not be affected until a branch is cut over. Data center sites can also act as transit for migrated branch sites to communicate with the rest of the network. A simple before/after scenario is shown in Figure 7.13.

The **Before** figure depicts a typical data center network with dual cores and WAN edges connecting to service provide MPLS network. The **After** figure depicts the connectivity between Mware SD-WAN Edges to the data center infrastructure.

The WAN links from the VMware SD-WAN Edge(s) are connected directly to a customer edge router for MPLS (private WAN access). The internet WAN links are connected to the firewall in the DMZ. The LAN links are connected to both data center core devices for redundancy.

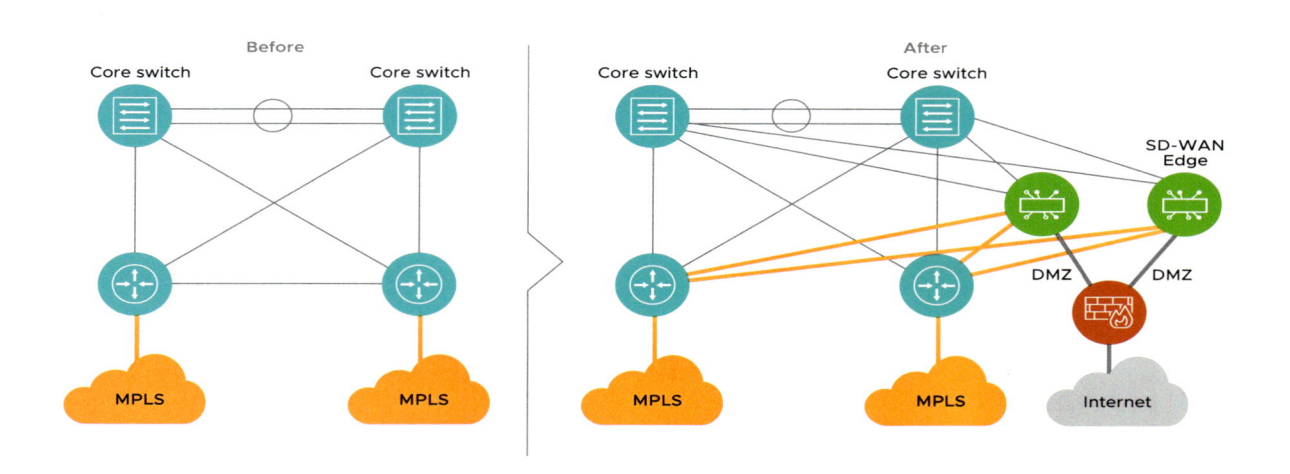

Figure 7.13: Data center insertion design

Depending on the requirements for throughput and scalability, the data center design can be deployed in a high availability (HA) configuration or in a cluster as discussed in Chapter 4: *Site Topology and Redundancy*.

> **TECH TIP**
>
> Clustering is suited for data center design as it allows horizontal scalability.

Data center insertion considerations

While building the data center insertion design, the following points should be considered:

1. Firewall rules: The following ports/protocols should be allowed on the firewall:

- TCP 443 (outbound)
- UDP 2426 (inbound & outbound)
- DNS (53) and NTP (123) (outbound)

2. WAN links: As shown in Figure 7.13, the WAN links from the VMware SD-WAN Edge(s) in the data center could be connected directly to the customer edge router for MPLS (private WAN access). Internet WAN links are connected to a firewall in a DMZ zone. Another alternative is to connect the WAN link from the VMware SD-WAN Edge to the data center core in the event that additional ports are not available on the MPLS customer edge routers.

> **TECH TIP**
>
> Automatic bandwidth measurement, i.e. Slow Start or Burst Mode, should not be used at the Hub site. This prevents large number of bandwidth detection tests from branch sites in the event the Hub edge is rebooted. Use "Do Not Measure (define manually)" instead.

3. Clustering: If VMware SD-WAN Edges are deployed in clustering, then the number of VMware SD-WAN Edges in the cluster should be carefully planned; such that in case of failure, the remaining number of VMware SD-WAN Edges in the cluster should be able to handle throughput and tunnel scale. The best practice is to plan for n+1 redundancy.

Step 3. Deploy VMware SD-WAN Edge at the Branch

Branch migration is the last step in the migration workflow. During branch migrations, production traffic is cut over from the traditional network onto the SD-WAN overlay network. All the steps before branch migration are seamless and non-intrusive to the existing network and require no downtime. Depending on the branch type (hybrid, MPLS-only, internet-only) as outlined in Chapter 4: *Site Topology and Redundancy*, different migration strategies can be followed. Figure 7.14 depicts different branch topologies and insertion designs.

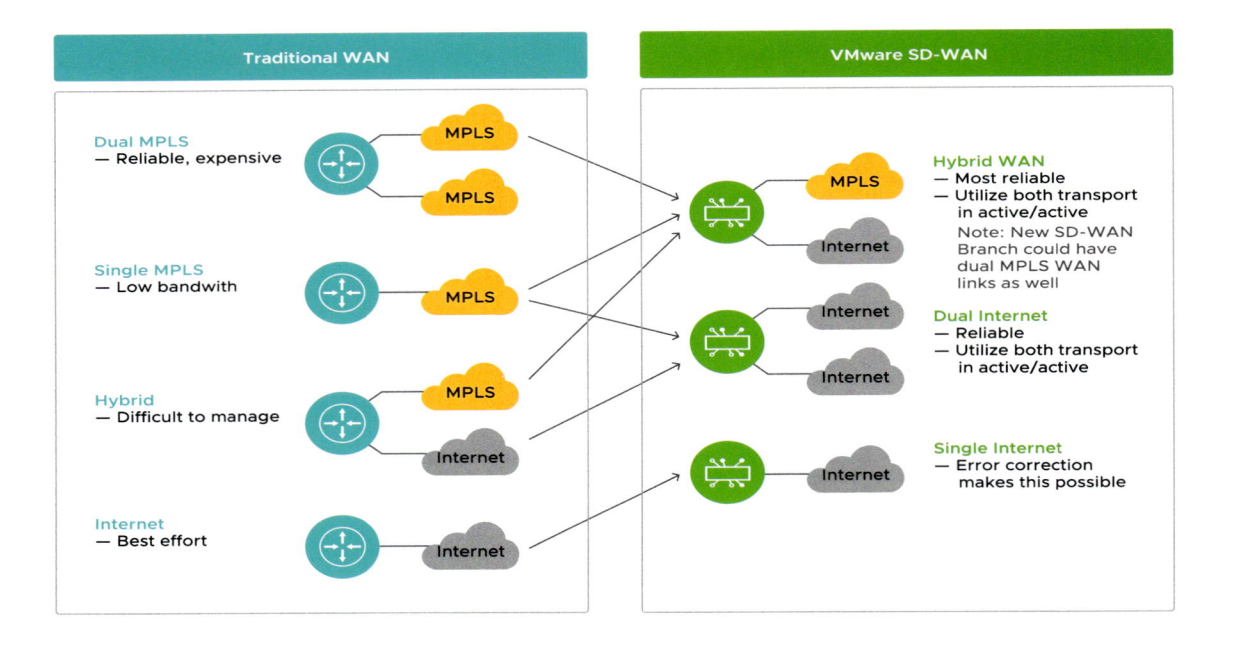

Figure 7.14: Branch Topologies and Insertion Types

Branch insertion considerations

While building the branch insertion design, the following points should be considered:

1. Firewall Rules: For those cases in which the firewall is placed in front of the VMware SD-WAN Edge, the following ports should be allowed:

- TCP 443 (outbound)
- UDP 2426 (outbound)
- DNS (53) and NTP (123) (outbound)

2. WAN links: Figure 7.15 depicts in-path design wherein WAN connections are terminated directly into the VMware SD-WAN edge.

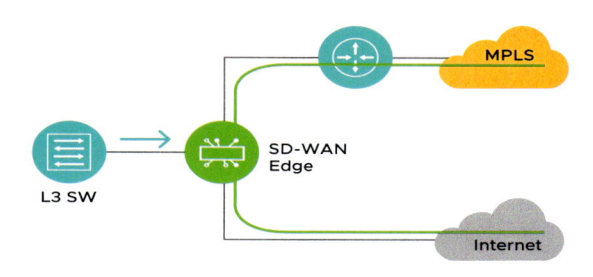

Figure 7.15: In-path design

Figure 7.16 depicts off-path design wherein one or more WAN links are not directly connected to the VMware SD-WAN Edge.

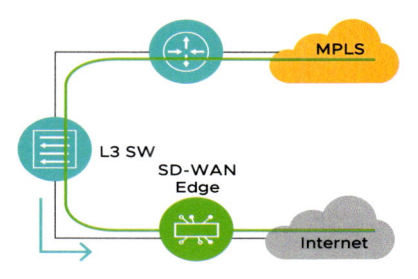

Figure 7.16: Off-path design

> **TECH TIP**
>
> WAN link bandwidth should not be shared with users not connected via the SD-WAN network. The VMware SD-WAN Edge should have complete visibility of WAN links for proper QoS and load balancing.

3. Routing/Traffic flows: The following routing/traffic flow requirements should be considered.

- WAN-side routing: If branch sites have no requirement to use underlay routing to reach other non-SD-WAN sites or applications, then no routing should be enabled on the WAN links of the branch VMware SD-WAN Edges. In the case where branch sites have requirements to use underlay routing (i.e. running BGP with a CE/PE) to reach other non-SD-WAN sites or reach any service provider offered services, the following two instances should be applied:

    - BGP uplink flag to CE/PE neighbor in case of in-path design.
    - BGP uplink community in case of off-path design.

Please refer to Chapter 5: *Routing* for an explanation of BGP uplink and BGP uplink community.

- Control plane reachability via private link: By default, VMware SD-WAN Edges use public links for control plane reachability to the VMware SD-WAN Orchestrator and VMware SD-WAN Gateways/Controllers. In order for VMware SD-WAN Edges to use MPLS to reach the cloud-hosted VMware SD-WAN Orchestrator and VMware SD-WAN Gateways, the `SD-WAN Service Reachable` as shown in Figure 7.17 should be enabled on all private links. There has to be a default route in the MPLS in underlay either advertised from the data center or by the service provider in order to facilitate this internet connectivity.

To enable control plane reachability via the private link on all branches, select `SD-WAN Service Reachable`. This feature can be enabled by navigating to `Configure Edge > Device > WAN Settings > Type User Defined > Select Edit under Actions`

User Defined WAN Overlay

Link Type:	Private
Name:	MPLS_Overlay
SD-WAN Service Reachable: ⓘ	✓

Figure 7.17: SD-WAN Service Reachable

4. Application QoS: By default, VMware SD-WAN Edge comes with standard business policies called `Smart Defaults` which classify and put applications into appropriate links and queues. Changing these default settings is not required in most instances. In cases where there are certain customer-specific applications, the application should be added to the respective business policies for the correct QoS classification and appropriate remediation techniques to be applied by the VMware SD-WAN Edge.

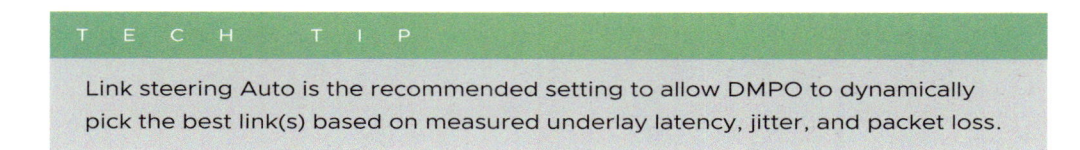

> **TECH TIP**
>
> Link steering Auto is the recommended setting to allow DMPO to dynamically pick the best link(s) based on measured underlay latency, jitter, and packet loss.

> **Key Takeaways:**
>
> - Introduce VMware SD-WAN Edge at the data center first
> - Remember to open the required ports/protocols at branch and data center
> - Enable SD-WAN Service Reachable for sites that have private overlays

Migration

SD-WAN Migration Journey

After another planning session with the VMware team, Rodney and Hans have a plan! Alvina believes it's time to get a quote for the VMware SD-WAN now that her team has a firm grasp of what they need to do and both Rodney and Hans seem to be oddly of a common mindset. Alvina is not sure that she has ever seen Hans and Rodney cooperating versus ready to jump into the octagon for a cage match. Could SD-WAN serve another purpose of bridging some of the gaps between her teams?

Rodney now understands the steps and the workflow to transform Allofu's existing network architecture to a fully software-defined WAN. With all the knowledge he has gained so far, Rodney comes up with a new design for Allofu's network architecture and builds guildelines, routing, and traffic flow requirements. Figure 7.18 below depicts Allofu's current network architecture.

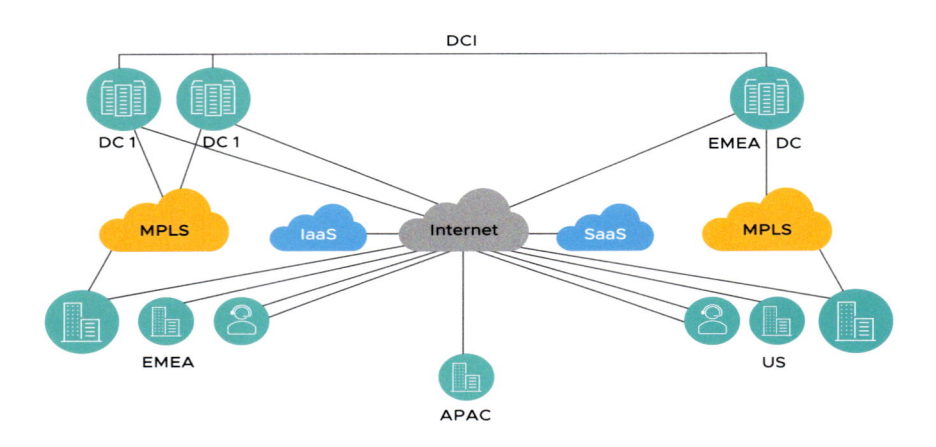

Figure 7.18: Allofu's network architecture

VPN Setup:

Branches and call centers within a region can build dynamic tunnels and will use regional hubs for initial tunnel build up. Inter-region dynamic tunnels will not be allowed. Rodney achieves this goal by assigning Edges to region profiles. For example, Figure 7.19 shows how the configuration will look for Allofu's US western regional branches.

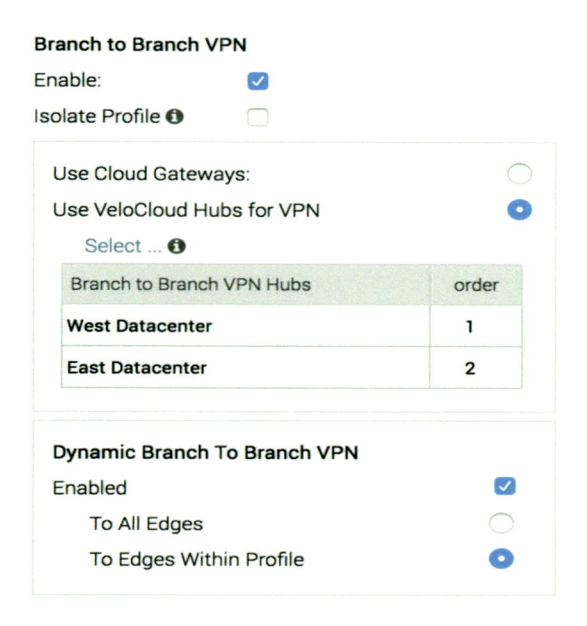

Figure 7.19: VPN Setup

Rodney loves the ease of configuration that profiles provide and decides to use only 3 profiles for the entire organization.

1. EMEA profile with EMEA DC as Hub.
2. US East region profile: US East DC as primary and US west DC as secondary.
3. US West region profile: US West DC as primary and US east DC as secondary.

Inter-region traffic:

Traffic from EMEA branches and call centers to the US region will flow to their respective regional Hubs and then use the Data Center Interconnect (DCI; provided by a third-party).

SD-WAN to non-SD-WAN traffic:

During migration, branches on the SD-WAN network would leverage the MPLS underlay to communicate directly with non-SD-WAN branches and use SD-WAN overlays to the regional hub as backup in case of local MPLS failure. This is a requirement due to Allofu's latency-sensitive ERP applications.

Asia Pacific branches:
Any new Asia sites will solely use the internet, connect to the EMEA hub, and will build dynamic tunnels directly to EMEA branches if required. In order for Asia branches to reach US data center or branches/call centers, traffic will flow through the EMEA data center and then over DCI to the US data center.

Acquisition:
If there is any acquisition by Allofu, the new company would be put in its own new segment. Using segmentation will make it very easy for Allofu to keep the traffic segregated and have it pass through a firewall before it reaches Allofu's infrastructure.

Local internet breakout:
All branches and call centers will have local internet breakout to reach SaaS and IaaS. In order to keep their branches secure, they will use Zscaler as a Cloud Access Security Broker.

Allofu's SD-WAN Infrastructure

After running a very successful pilot, Rodney decides that VMware hosted SD-WAN Orchestrator and Gateways are the best solutions for his enterprise.

Data center deployment

Rodney tested clustering in the pilot and is convinced that it is the best solution for the data center architecture as it scales horizontally with respect to tunnels and throughput. Rodney asks his security team to open required ports on the firewall to allow the VMware SD-WAN data center to connect to the VMware SD-WAN Orchestrator and Gateways. He decides to keep the off-path design built already in the pilot for all its data centers during migration. Using this approach, he doesn't impact any of his existing infrastructure and he doesn't have to schedule any downtime.

As per Rodney's routing requirements, the data center would act as a backup path for SD-WAN branches to communicate with yet to be migrated branches. In order for this to work and not cause any routing issues, Rodney comes up with following requirements:

1. Community for SD-WAN routes: 12345:888
2. Community for MPLS and DCI routes: 12345:999
3. Regional data centers will do AS-Path prepend for the SD-WAN branches/call center routes before advertising it into MPLS as it will help SD-WAN branches to still have connectivity to the rest of the non-SD-WAN branches in case they lose their MPLS link.

Rodney is delighted that by just configuring a community on the CE router and having that as the Uplink Community on the BGP neighbor, it became very easy for SD-WAN branches and call centers to use their data centers as backup to reach non-SD-WAN branches during migration.

Branch migration

From the pilot, Rodney has learned all the different branch topologies and insertion strategies and has developed the following migration plan for branch and call centers.

1. Large Branches: One MPLS and one internet link at large branches as VMware SD-WAN DMPO technology provides the same level or better performance on internet links during pilot phase. Since these sites are large and critical to the business, VMware SD-WAN edges will be deployed in HA mode.
2. Smaller branch: Single internet links and no HA.
3. Call centers: Running call centers purely on dual internet during pilot proved very successful so all call centers will have dual internet connections and will be deployed in HA topology.

Since branch locations will be running BGP with the MPLS Provider routers during migration, every branch will mark these BGP neighbors as uplink to avoid any branch locations from becoming a transit site.

Rodney is super delighted with this plan and looks forward to another great session of goat yoga later that day!

8. Service Provider

Given the overriding cost sensitivity of Allofu, Alvina continues to wrack her brain for ways to improve business efficiency and reduce cost. Allofu has always designed, deployed, and maintained its own infrastructure, but in light of industry trends surrounding the cloud, is this a good time to consider the possibility of purchasing SD-WAN as a service from a managed service provider? Rodney and his team would likely be resistant as the WAN is such an integral part of their role today, and she even heard Rodney refer to the WAN as his 'baby' the other day. It is also likely that Rodney may be a bit resistant to the hosted solution as it may appear to be career threatening to him. In truth, Alvina values Rodney's contributions and has a number of other strategic projects she would gladly steer his focus to, were he not spending so much of his time on day-to-day tactical activities supporting the WAN/branch infrastructure. Alvina thinks it's time to explore some options with Circuits 'R' Us and she has recently learned from her account representative that they, in fact, have a managed service offering utilizing VMware SD-WAN.

Service Provider Topology

Service Provider Deployment Model (Provider-hosted or On-premises Deployment)

In a service provider environment, the VMware SD-WAN Gateway is able to be deployed as a partner gateway, such that the partner gateway is connecting to both the internet and service provider's MPLS backbone. Figure 8.1 shows a typical service provider architecture.

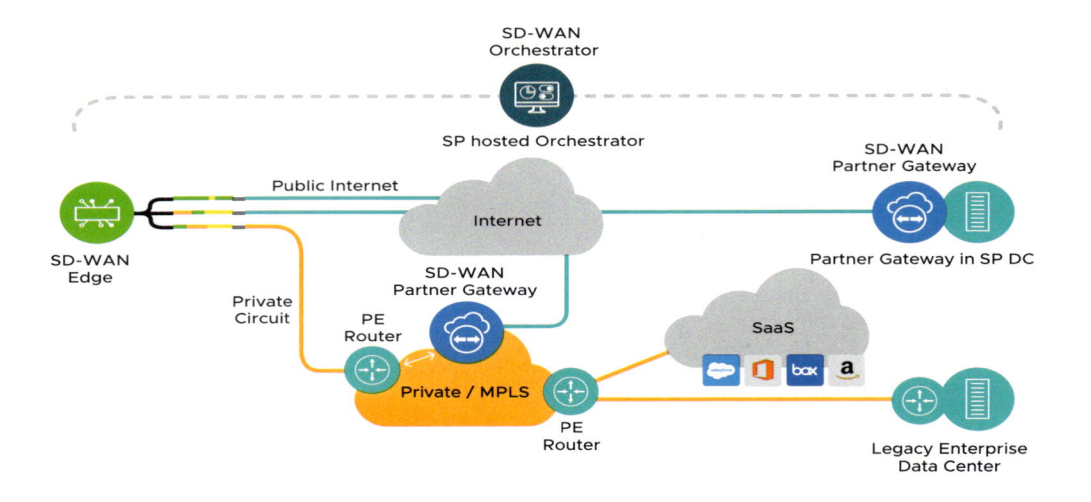

Figure 8.1 - VMware SD-WAN overall architecture for the service provider-hosted offering

One of the main advantages of the service provider-hosted model is in the function of the gateway itself. This partner gateway functionality provides an advantage to the service provider in hosting the gateway themselves.

Partner Gateway

Multi-tenant VMware SD-WAN Gateways are deployed at service provider points-of-presence (PoP) to provide secure handoffs segmented by customer, and segmented within the customer to the applications or networks.

There is no difference in software between a VMware SD-WAN Partner Gateway and a VMware hosted SD-WAN Cloud Gateway. Instead, the difference comes with the allocation of the SD-WAN Partner Gateway role to the gateway itself, which provides additional functionality on top of the standard cloud-hosted gateway configuration.

This functionality allows for handing off the traffic in a per-enterprise and per-segment fashion to the second interface allocated on the gateway towards the PE/DC router. This handoff is covered in more detail in Chapter 8: *Service Provider Routing*.

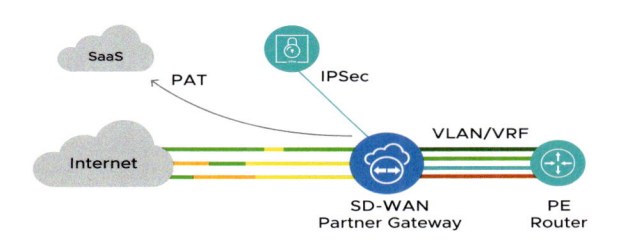

Figure 8.2: SD-WAN partner gateway

This functionality allows for differentiated and secure service delivery via the same partner gateway, allowing for rapid deployment and scaling of the infrastructure.

The other defining characteristic of the partner gateway is that the allocation to an SD-WAN edge is deterministic. This allocation can be performed at the profile level or at an edge level override. This deterministic assignment is a critical factor in helping to ensure that the proper gateways are assigned to SD-WAN edges, based on the associated services being provided from that gateway to the end customer.

MPLS Integration Deployment Use Case for a Service Provider

VMware SD-WAN adapts to the service provider network architecture to accommodate the shift of applications to the cloud and to increase the available bandwidth at a lower cost while maintaining the expected levels of reliability, performance, and security. By offering VMware SD-WAN as a service, the service provider can protect its MPLS network investment by augmenting it with other connection types. In addition, the service provider can add value for their customers by providing a more agile and better-performing network. This also allows a service provider to extend beyond its typical network footprint and reach a broader customer base.

Provider Services via the Overlay

The partner gateway provides a way to give access to other service offerings that the service provider has as part of their current product set, such as voice, PCI environments, cloud firewall, and public IP backhaul. Each of these offerings would be front-ended by a partner gateway sitting at the associated service center and advertising reachability of those services to any edges attached to that partner gateway. The prescribed partner gateway deployment would follow the same redundancy strategy, with the ideal scenario allocating geo-redundant gateways to mitigate any outages that may impact an entire service center. In scenarios where geo-redundancy is not possible or practical, redundancy in the same service center is still recommended, with partner gateways deployed on discrete compute stacks. Figure 8.3 highlights a triple play service use case where the same gateway is offering access to an individual customer with VRF separated handoffs to each service offering.

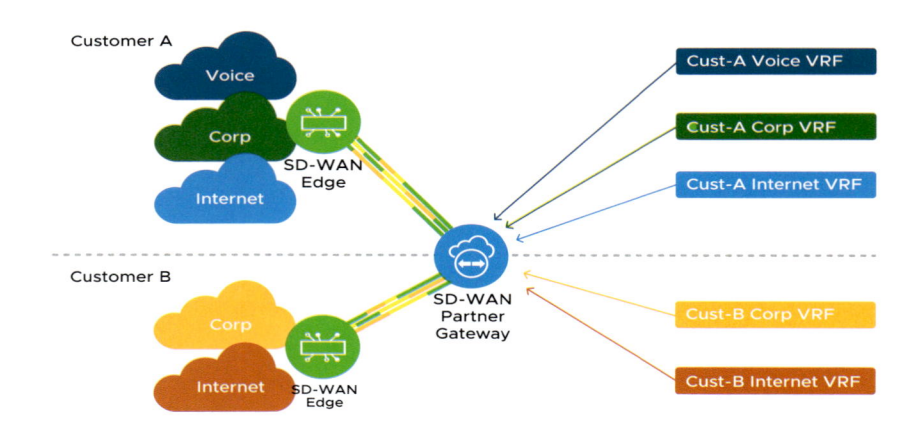

Figure 8.3: Accessing provider services via the SD-WAN partner gateway

Inbound Public IP backhaul

There is also the use case where an enterprise wishes to host an inbound service available to outside users such as a website or VPN connectivity in a highly available fashion. In such use cases, the service provider can allocate a public IP from their pool to be placed on the LAN side of the edge and provided at the customer site. Inbound requests for this IP will hit the service provider internet edge, be routed through their core to the PE for the appropriate gateway and forwarded on to the edge and the client as depicted in Figure 8.4. This topology allows the service to remain available regardless of what transport is connected to the edge, even when it is not owned by the service provider offering the public IP.

A few key points regarding this inbound public IP backhaul:

- The customer has their own public IP block assigned by SP, does not want NAT
- Multiple links provide aggregation, resiliency while retaining customer IP block
- SP optionally provides security for service for internet traffic
- All traffic is forced to go through SP first
- If public IP is not a requirement, can optionally assign private block.

Figure 8.4 illustrates the topology.

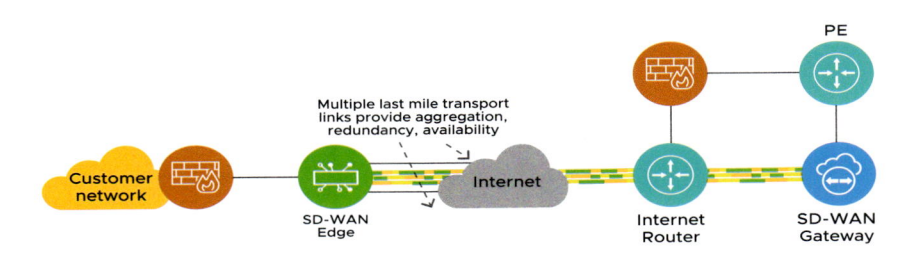

Figure 8.4: Inbound public IP address backhaul

Service Provider 183

Shared Services via Policy-based NAT

There are times where a service provider may offer a common service to many customers that is not natively VRF-aware. In those cases, the possibility of overlapping subnets between customers exists, and it becomes necessary to perform a policy-based NAT to facilitate service offerings without having to worry about address collision. The policy NAT is performed on the partner gateway handoff interface directly prior to being sent on to its destination via the PE. This allows the shared resource to return the traffic back to the gateway that originates the traffic which will then be forwarded to the appropriate customer and edge. In the example following, a source policy NAT could be defined to allow for traffic coming from Customer A - 10.0.0.0/24 to translate to 1.1.1.1/32 and from Customer B - 10.0.0.0/24 to translate to 2.2.2.2/32.

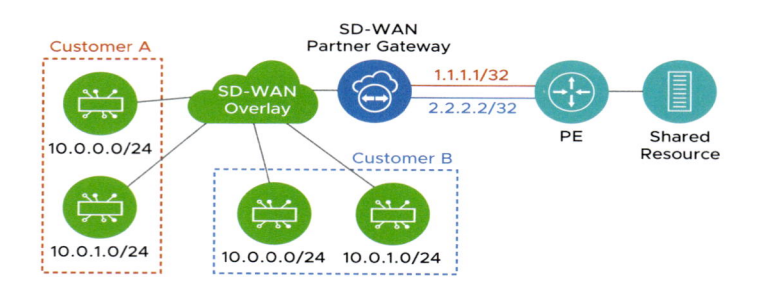

Figure 8.5: Shared service via policy-based NAT

Mid-mile Transport

Many service providers operate extensive inter-region networks that have tightly controlled SLAs. In customer deployments that span multiple regions, the service provider can selectively route this traffic via region-specific partner gateways to then be carried across the mid-mile transport towards partner gateways in another region. This helps to provide a path to mitigate potential long-haul connectivity issues that might otherwise be present in a traditional internet-only connectivity scenario. Figure 8.6 shows a high-level view of this use case:

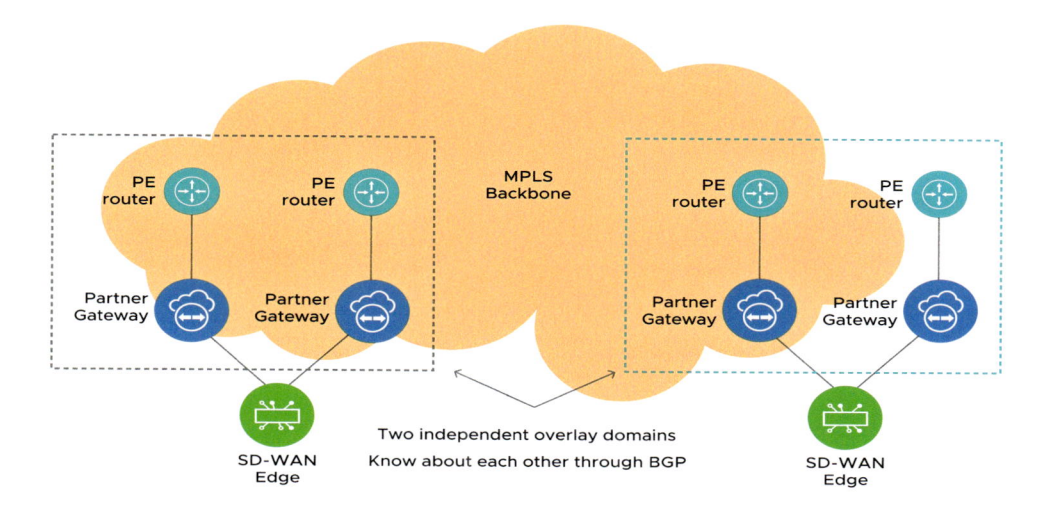

Figure 8.6: Mid-mile transport with VMware SD-WAN partner gateway

This scenario is also often called last-mile replacement. The VMware SD-WAN Edge is able to use the internet as the last mile to reach the partner gateway as an entry point to the service provider's MPLS backbone, which makes the private local loop no longer mandatory at the SD-WAN Edge.

> **Key Takeaways:**
>
> - Fundamentally, it is the same software for SD-WAN Partner Gateway and VMware SD-WAN Gateway. The difference comes from the role each plays in the network.
> - Partner Gateway is hosted on the premises of the service provider or enterprise customer. It offers secure hand-off of traffic on a per-enterprise or per-segment basis to PE/DC router, and allows a service provider to offer the following differentiated services:
>
>   - Provider services via SD-WAN Overlay
>   - Inbound internet backhaul
>   - Shared services via policy-based NAT
>   - Mid-mile transport

Service Provider

Service Provider Routing

Partner Gateway Assignment

In the Over The Top model, VMware SD-WAN Gateway assignment to edges is performed using geo-location. In the service provider deployment model, partner gateways are statically assigned by the service provider. Generally, a primary and secondary partner gateway in the same region will be assigned to the VMware SD-WAN Edge. The static assignment of partner gateways allows the service provider network engineer to control traffic flow between VMware SD-WAN Edges and to provider services.

Traffic Path with Mid-mile Transport and Partner Gateway

It is important to understand the traffic path taken by the traffic from one VMware SD-WAN Edge to the other VMware SD-WAN Edge when the traffic is passing through the partner gateway and MPLS backbone.

In Figure 8.7, the VMware SD-WAN Edge and associated partner gateways on the left-hand side are located in region 1. The VMware SD-WAN Edge and associated partner gateways on the right-hand side are located in region 2.

When a client on the LAN side of VMware SD-WAN **Edge11** sends packets to VMware SD-WAN **Edge21**, it does so in 3 steps.

- First, the packets will be sent from VMware SD-WAN **Edge11** to partner gateway **GW11** over the SD-WAN Overlay.
- Second, the packets will be routed from **GW11** in the MPLS backbone via **PE11** to reach partner gateway **GW21** via **PE21**.
- Third, the packets will be sent from partner gateway **GW21** to the VMware SD-WAN **Edge21** over the SD-WAN Overlay.

One important point to note here is the region 1 and region 2 partner gateways do not form any SD-WAN Overlay between each other. This reachability is achieved by BGP routing on the backbone.

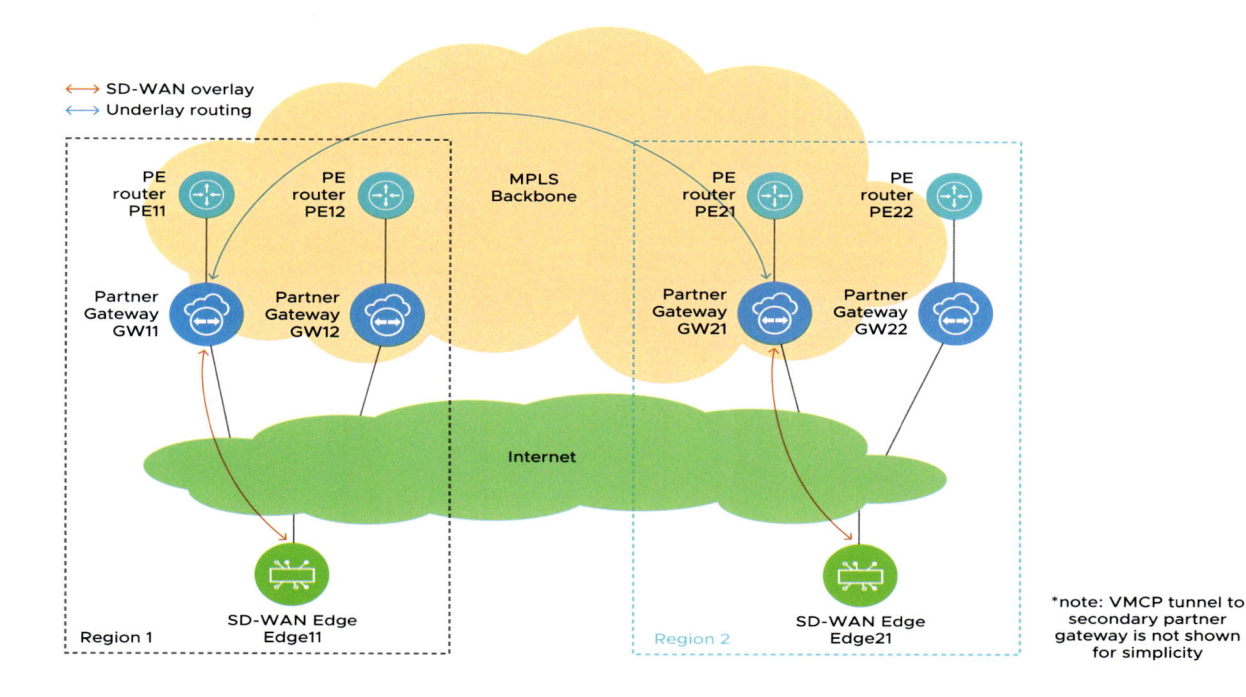

Figure 8.7: Traffic path between two VMware SD-WAN Edges located in two different regions

As previously mentioned, the gateway is multi-tenanted and able to connect to multiple VRFs with the PE. Each VRF is identified by either a single VLAN tag (802.1q) or double VLAN tag (QinQ). Optionally, no tag can be applied for use cases where an explicit per customer VRF handoff is not needed. This is a highly scalable architecture, such that capacity can be horizontally scaled out by adding more gateways when necessary.

Traffic Path between SD-WAN and Traditional Branches

The architecture also allows compatibility with traditional sites that only have MPLS connections. Figure 8.8 illustrates the condition when there is traditional router connected to the customer VRF.

In the sample scenario, a branch with a CE router **CE21** is connected to the service provider MPLS backbone. Router **CE21** is advertising a prefix L1 which is the LAN side subnet. Prefix L1 is advertised via BGP to the PEs in the customer VRF. Due to the partner gateways using BGP to peer with the PE routers, they also learn **L1** and can advertise it to the connected VMware SD-WAN Edges. As a result each VMware SD-WAN Edge is able to reach the traditional MPLS site via their partner gateways. This is useful during the migration or in some situations where a particular site is not able to migrate to SD-WAN.

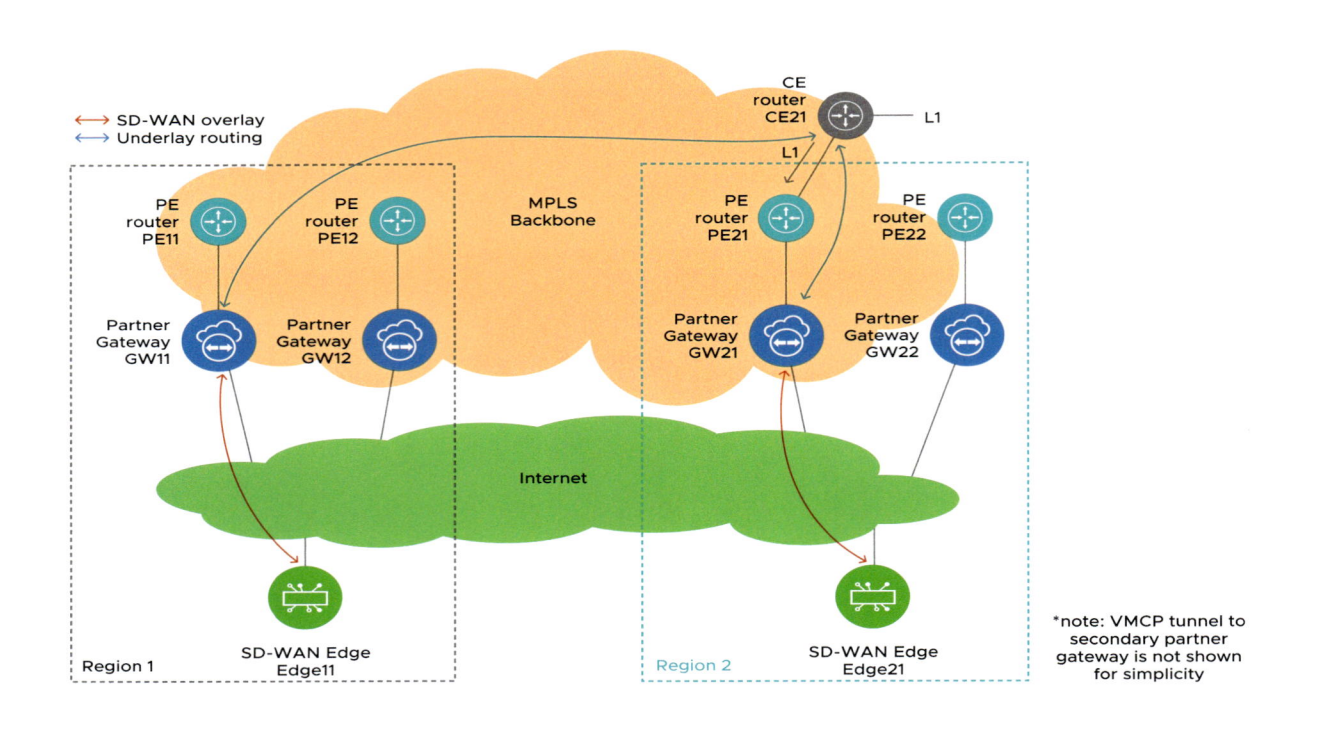

Figure 8.8: Using SD-WAN partner gateway to connect with a traditional MPLS site

Traffic Path between VMware SD-WAN Edges with Common Partner Gateway

The decision on whether the traffic is direct between the two VMware SD-WAN Edges or if traffic has to go through the common gateway, will depend on the CloudVPN configuration which was referenced earlier in Chapter 4: *Overlay Topologies*. If there is at least one common partner gateway assigned for two VMware SD-WAN Edges, the traffic between these two VMware SD-WAN Edges will not enter the MPLS backbone. Instead the traffic flows directly between the two VMware SD-WAN Edges or via the partner gateways using the SD-WAN overlay.

- Dynamic Branch to Branch VPN is enabled:", that is become "Dynamic Branch to Branch VPN is enabled: It will flow between VMware SD-WAN Edge11 to VMware SD-WAN Edge12 via direct SD-WAN overlay. It will flow between VMware SD-WAN **Edge11** to VMware SD-WAN **Edge12** via direct SD-WAN overlay.

- Dynamic Branch to Branch VPN is disabled:" that is become "Dynamic Branch to Branch VPN is disabled: It will flow from a VMware SD-WAN Edge11 to partner gateway GW11 then to VMware SD-WAN Edge12. It will flow from a VMware SD-WAN **Edge11** partner gateway **GW11** to a VMware SD-WAN **Edge12**.

Service Provider

Figure 8.9 demonstrates how the traffic flow is affected by the partner gateway assignment:

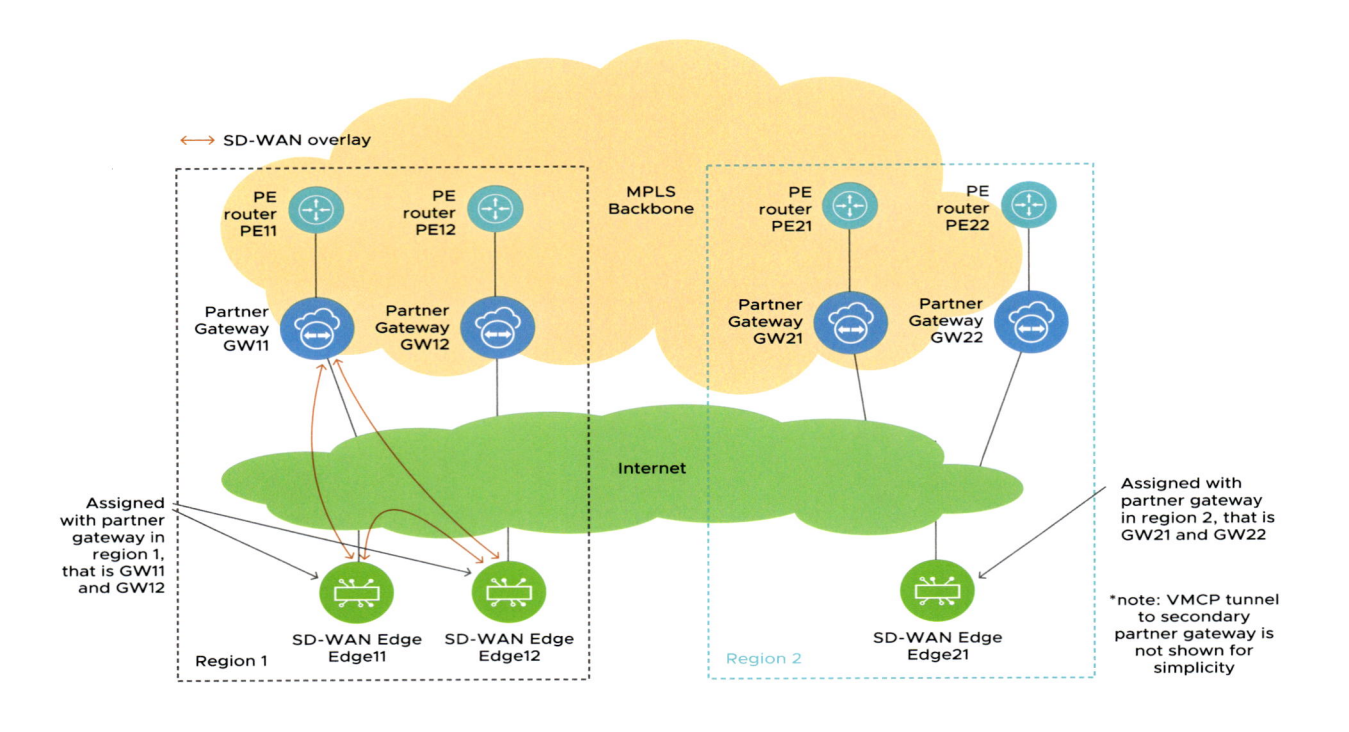

Figure 8.9: Different partner gateway assignment scenario

SD-WAN Overlay over the MPLS Network

Another advantage of the partner gateway is that a VMware SD-WAN Edge with both internet connectivity and MPLS connectivity from the service provider hosting the gateway can have VCMP tunnels terminate on both the public and private side of the gateway. Figure 8.10 illustrates this scenario.

In this sample figure, **Edge11** has both internet and private MPLS underlay connectivity. **Edge11** forms a VCMP tunnel with **GW11** via both underlay connections. This introduces some additional benefits as follows:

- Traffic between the VMware SD-WAN Edge and partner gateway can benefit from DMPO using both the internet and private MPLS circuits.

- In the event that the VMware SD-WAN Edge loses internet connectivity, the VMware SD-WAN Edge is still manageable on the VMware SD-WAN Orchestrator. The management plane traffic is able to go through the SD-WAN overlay on the private MPLS side to reach the VMware SD-WAN Orchestrator through the partner gateway.

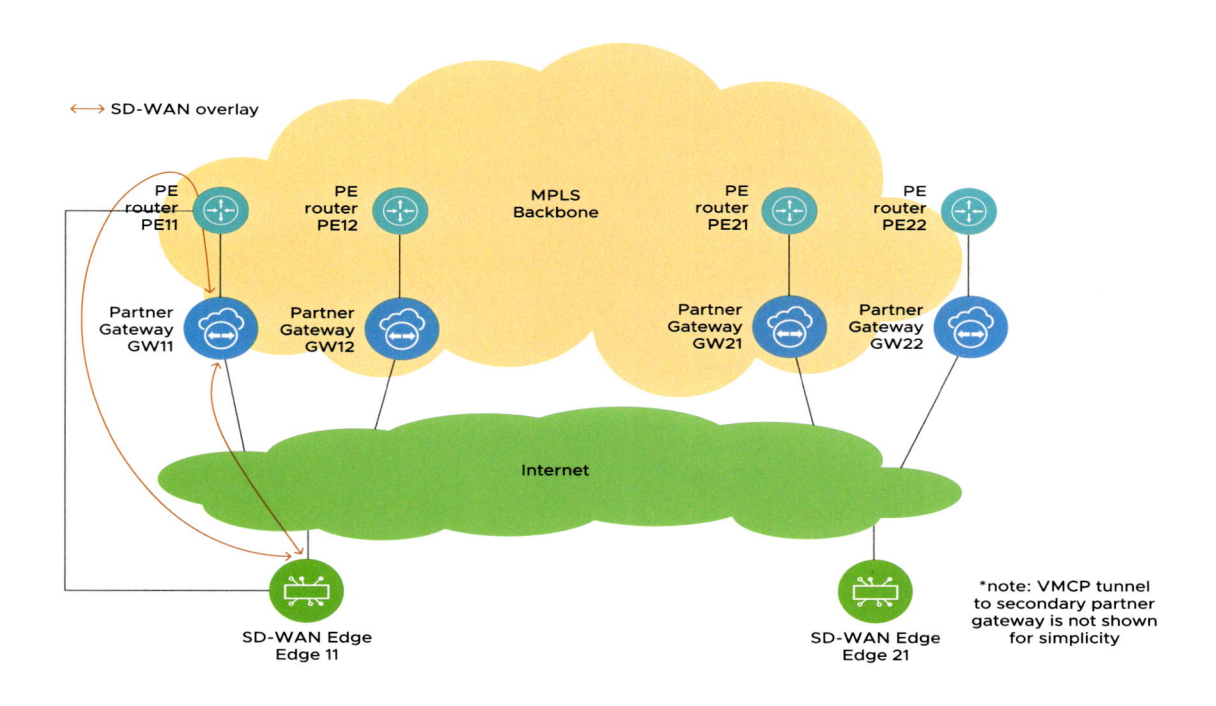

Figure 8.10: VMware SD-WAN Edge forms SD-WAN Overlay with SD-WAN partner gateway on the private MPLS

VMware SD-WAN Partner Gateway Order

When assigning the partner gateway to the VMware SD-WAN Edge, the manual assignment requires an order number to be selected. The screen capture in Figure

Service Provider

8.11 shows **GW11** with an order number 1, while **GW12** has an order number 2. The smaller the number, the higher priority it is.

There are a few points to highlight for the meaning of the order number:

- When the VMware SD-WAN Edge learns prefixes from the partner gateway, the original BGP attributes such as AS-PATH metrics are retained. The VMware SD-WAN Edge will select the preferred partner gateway based on the prefix attribute. If the prefix attribute is identical, the order number will be the tiebreaker, that means the partner gateway with order 1 will be used.

- When the partner gateway is as shown Figure 8.9, the VMware SD-WAN Edge which initiates the traffic will always send the traffic to the partner gateway with order 1 for itself unless the SD-WAN overlay to that partner gateway is down.

- The partner gateway advertises prefixes learned from the VMware SD-WAN Edges to the PE router. During this route advertisement, the order number becomes the metric. This means that partner gateway with order 1 will advertise the prefix with metric 1 to the PE. Partner gateway with order 2 will advertise the route with metric 2 to the PE. This is trying to influence the PE routers such that the partner gateway with order 1 is the preferred partner gateway to reach the corresponding prefix.

Navigate to `Configure > Profile > {profile name} > Device tab > Edit Gateway Handoff Assignment`

Gateway Handoff Assignment

Gateways	Order
GW11	1
GW12	2

Select Gateways

Figure 8.11: VMware SD-WAN partner gateway order number

> **TECH TIP**
>
> It is not recommended to assign all partner gateways to every VMware SD-WAN Edge available in an enterprise. The SD-WAN Overlay from the VMware SD-WAN Edge to the partner gateway is persistent. Assigning an excessive number of partner gateways to the VMware SD-WAN Edge will result in a large number of SD-WAN Overlays being established in the SD-WAN Edge but not in use, which creates unnecessary overhead.

Symmetric Route with Partner Gateway

Pertaining to the previous section, the partner gateway uses the order number as the metric when advertising the route to the PE router. However, in a situation with two partner gateways using a different AS number, the PE router might not consider the metric number for route selection. This can result in an asymmetric path situation, as seen in Figure 8.12.

When the VMware SD-WAN **Edge11** sends traffic to VMware SD-WAN **Edge21**, the prefixes from VMware SD-WAN **Edge21** get the same attribute from **GW11** and **GW22**, and VMware SD-WAN **Edge11** will send the traffic to **GW11** as it is the partner gateway with order 1.

The PE routers will likely ignore the metric when **GW11** and **GW12** are using different AS numbers, and the PE routers are not comparing metrics from different AS. The return traffic from VMware SD-WAN **Edge21** may come to **GW12**, which will then forward the return traffic to VMware SD-WAN **Edge11**. The connectivity between VMware SD-WAN **Edge11** and VMware SD-WAN **Edge21** is successful but undesirable because the traffic path is asymmetric.

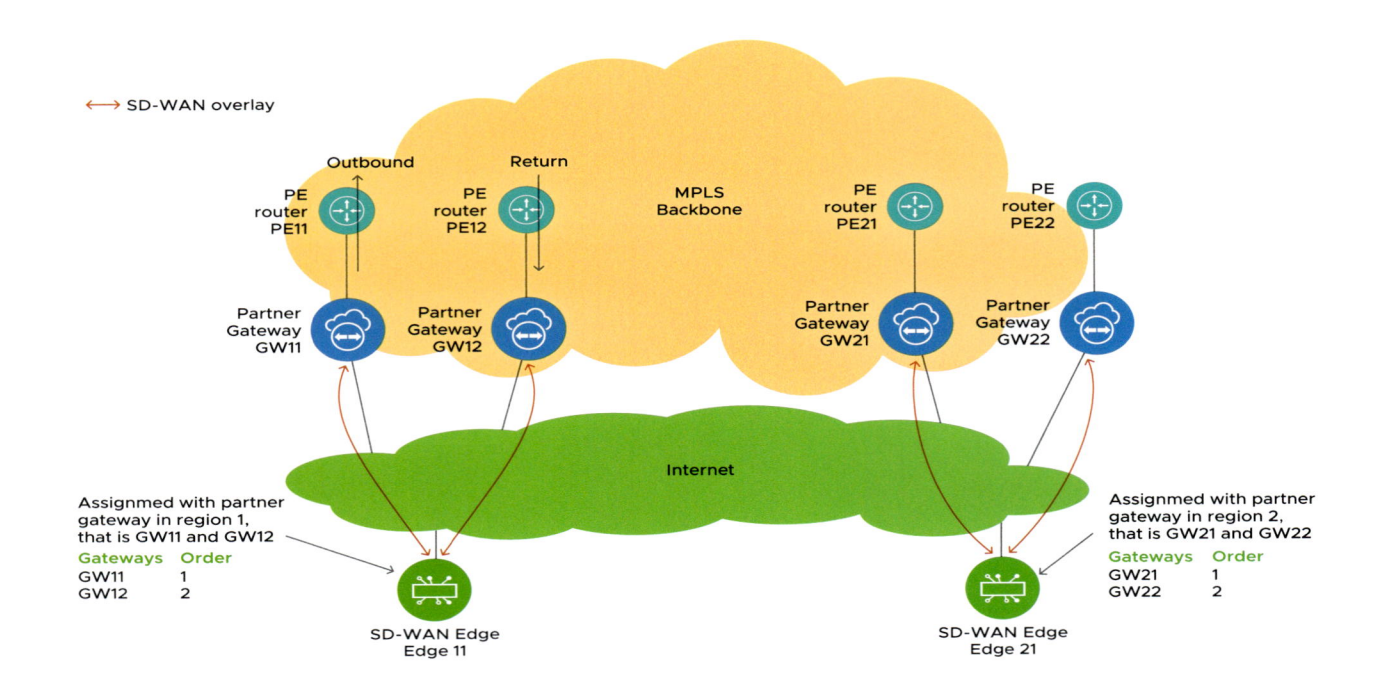

Figure 8.12: Undesirable asymmetric situation

To avoid this asymmetric situation, it is recommended to utilize the BGP community mapping feature.

Figure 8.13 shows a sample BGP community mapping configuration. In this sample, the partner gateway with order 1 will attach BGP community **65000:100** to all the prefixes it advertises to the PE router. For partner gateway with order 2, it will attach BGP community **65000:90** to all the prefixes it advertises to the PE router.

In the PE router, there will be a route-map to set a local preference of 100 when the prefix is received with a BGP community of **65000:100** and set a local preference of 90 when the prefix is received with a BGP community of **65000:90**. With the BGP community mapping set and corresponding route-map in the PE routers, asymmetric paths across different partner gateways will be avoided. Navigate to `Configure > Customer > Configure Hand Off > Customer BGP Priority > Enable Community Mapping`

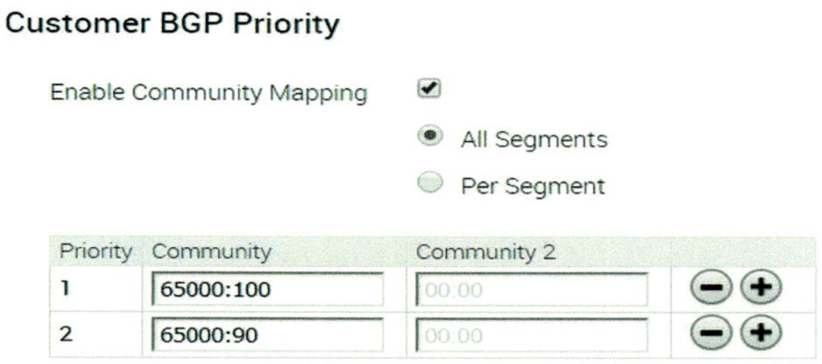

Figure 8.13: Auto BGP community mapping

Default Route from Partner Gateway

In cases of a gateway where partner handoff has been turned off, the gateway always advertises a default route to the connected VMware SD-WAN Edge. In the partner gateway scenario, the service provider administrator has an option to decide whether to have the partner gateway advertise a default route or not, but it is not advertised by default. A default can be configured as a static route and set to NAT on the gateway. This will allow traffic in that enterprise to use the public handoff interface of that Gateway to send traffic out to the internet, emulating the behavior of a standard VMware SD-WAN Gateway.

Key Takeaways:

- Partner gateways do not form SD-WAN Overlays amongst themselves. If traffic from one partner gateway needs to reach another partner gateway in the MPLS backbone, this is based on BGP routing.
- The assignment of partner gateway to the VMware SD-WAN Edges governs what the traffic flow will be between the VMware SD-WAN Edges.
- It is possible and recommended to form a private SD-WAN Overlay from the VMware SD-WAN Edge to the partner gateway if the VMware SD-WAN Edge comes with a private MPLS circuit connected to the MPLS backbone.
- Measures should be taken to ensure the traffic going out from the partner gateway to the MPLS backbone return on the same partner gateway to avoid asymmetric routing.

Alvina is very interested in the managed service offering from Circuits 'R' Us, but pricing is key and she really needs to see the quote before she can begin to explore next steps. Circuits 'R' Us has also chosen the VMware SD-WAN solution, which is encouraging given Allofu's own research into the solution. Circuits 'R' Us has plans to offer additional services natively in their private cloud to encourage Allofu to use their managed service offering. Alvina will wait for a quote and then decide, but both are viable options!

9. Integration

Alvina recalls a few 'palm to forehead' moments during some previous network cutovers where critical areas, like troubleshooting and debugging tools and process, were lost in the rush to implementation. Alvina also wants to ensure that Rodney and his team are prepared to not only deploy an SD-WAN network but to also operationalize and troubleshoot it. Alvina is encouraged, based on Rodney's earliest experiences with the VMware SD-WAN tools for traffic visibility, but needs to get confirmation from Rodney, Hans, and the Network Operations Center (NOC) team. Alvina is curious to see if there is a way for her to develop a clear ROI that she could share with the Board of Directors at the next board meeting?

Rodney knows that the Allofu NOC is heavily invested and reliant upon their NMS. He needs to ensure that the NOC makes minimal changes to accommodate SD-WAN in the Allofu WAN. So far, the tools have been easy to use, but it's time to make sure the NOC team is in agreement as well and Rodney invites them to join a web conference with the VMware team to walk the team through the toolset. Rodney senses that the finish line may actually be visible, but this is a crucial final hurdle and Rodney does recall how painful missing a hurdle can be from his time in middle school track.

Operations

A critical, if not mundane, aspect of any network infrastructure is the operational side of things. When undertaking the journey to move to a SD-WAN centric topology, there are considerations for both existing and new operational best practices to maintain the status quo and ensure any potential disruptions are minimized.

Network Management

On the operational front, the VMware SD-WAN Orchestrator contains a wealth of information that is readily accessible to the network engineer. Among the simple and straightforward items are device health checks around system utilisation and health.

More detailed information can be gleaned regarding individual application utilization statistics: who is using those applications; how applications are flowing through the network; as well as insight into the health of individual links from a latency, loss, and jitter perspective. There is also insight into the overall global routing table per enterprise that details where prefixes are learned and allows for easy manipulation of those individual prefixes.

Event and audit logs are available, such as configuration related data, which user executed that configuration change, and what the extent of that change was. To this end, an automatic configuration rollback is available on the VMware SD-WAN Edge itself. In case a configuration error makes the VMware SD-WAN Edge lose its connection to the VMware SD-WAN Orchestrator, the VMware SD-WAN Edge will try for 5 minutes to get the connection established. After this window of lost connectivity, the VMware SD-WAN Edge in question restores to its last known good configuration automatically to regain connectivity to the VMware SD-WAN Orchestrator and generates an event indicating that a bad configuration was pushed. Lastly, there are configurable events that can trigger SNMP traps, email or text-based alerts, or any combination of those items to stay abreast of any network impacting events.

The following provides details about the troubleshooting tools available to a network engineer in the VMware SD-WAN Orchestrator to further aid in managing operational efficiency.

Network Management Protocols

To integrate into an existing management environment, VMware SD-WAN offers industry-standard network management interfaces that can deliver additional insights into the network.

The most commonly used is SNMP which can be utilized to gain health and throughput statistics directly from the VMware SD-WAN Edge.

SNMP needs to be explicitly configured on the VMware SD-WAN Edge itself or the corresponding profile.

Additionally it needs to be allowed in the Firewall section (Edge or Profile possible).

Figure 9.1 and Figure 9.2 show a simple configuration for `SNMPv2`. Navigate to `Configure > Edge > {edge name} > Device tab > SNMP`

SNMP Settings

- Versions Enabled: ☑ v2c ☐ v3
- * Port: 161
- SNMP v2c Config
- * Community: YourCommunityh3r3
- Allowed IPs: ☐ Any
  - 10.10.10.10

Figure 9.1: SNMP configuration

Navigate to `Configure > Edge > {edge name} > Firewall tab > SNMP Access`

SNMP Access

○ Deny All
○ Allow All LAN
● Allow the following IPs
  - 10.10.10.10

Separate each IP with a comma (,)

Figure 9.2: SNMP firewall settings

Integration

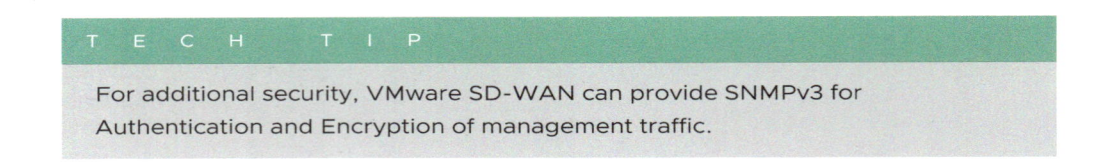

> **TECH TIP**
>
> For additional security, VMware SD-WAN can provide SNMPv3 for Authentication and Encryption of management traffic.

Additional flow statistics can be exported via NetFlow v10 to flow collectors for further processing. A typical deployment of NetFlow has branches exporting flows to a central collector. Common use cases for this type of flow-aggregation are performance and security reporting tools. Navigate **Configure > Edge > {edge name} > Device tab > NetFlow**

```
Netflow Settings
    Netflow Enabled: ☑
    Version: v10
    Collector IP:    [Ex: 54.183.9.192]
    Collector Port:  [4739]
```

Figure 9.3: NetFlow configuration

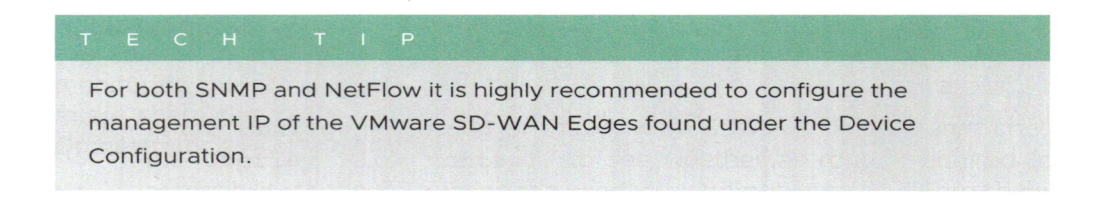

> **TECH TIP**
>
> For both SNMP and NetFlow it is highly recommended to configure the management IP of the VMware SD-WAN Edges found under the Device Configuration.

Management Automation

An additional operational method that bridges the gap between new and traditional operational consumption that leverages internal components is the REST API. Further details on utilizing the API are available in Chapter 9: *API*, however this is a powerful tool that can be used to the extent of even replacing the front-end UI of the orchestrator itself via a middleware platform should that be desired. Using the REST API, methods can be developed to roll out configuration changes that touch hundreds or even thousands of devices, or poll for information across an entire network to quickly correlate data during a network impacting event.

Key Takeaways:

- The VMware SD-WAN solution offers various tools to aid the simple management principle.
- There is a three-tier role-based access control system to enable different levels of operation for multi-tenancy and to support provider and customer level logins with restricted views to their environments.
- Standard network management protocols like IPFIX and SNMP are available.
- Additionally, the whole solution can be automated by using REST APIs with the VMware SD-WAN Orchestrator

Troubleshooting

VMware SD-WAN Orchestrator Troubleshooting Tools Overview

This chapter is a hands-on overview of the troubleshooting tools available with the VMware SD-WAN Orchestrator for a network engineer with customer level access to the VMware SD-WAN Orchestrator. Next to the `Monitor` section in the UI, there are four different sections for specific troubleshooting tools which provide different levels of information. The first three of them are found in the `Test & Troubleshoot` menu on the left-hand side of the UI as shown in Figure 9.4. Navigate to `Test & Troubleshoot > Remote Diagnostics`.

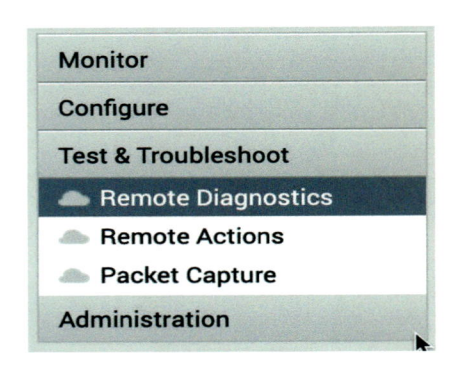

Figure 9.4: Test & Troubleshoot Menu

Remote Diagnostics

The `Remote Diagnostics` area is a key component in the `Test & Troubleshoot` menu. This collection of commands are used to display various states of information available in the VMware SD-WAN Edge. Some of the available commands are shown in Figure 9.5.

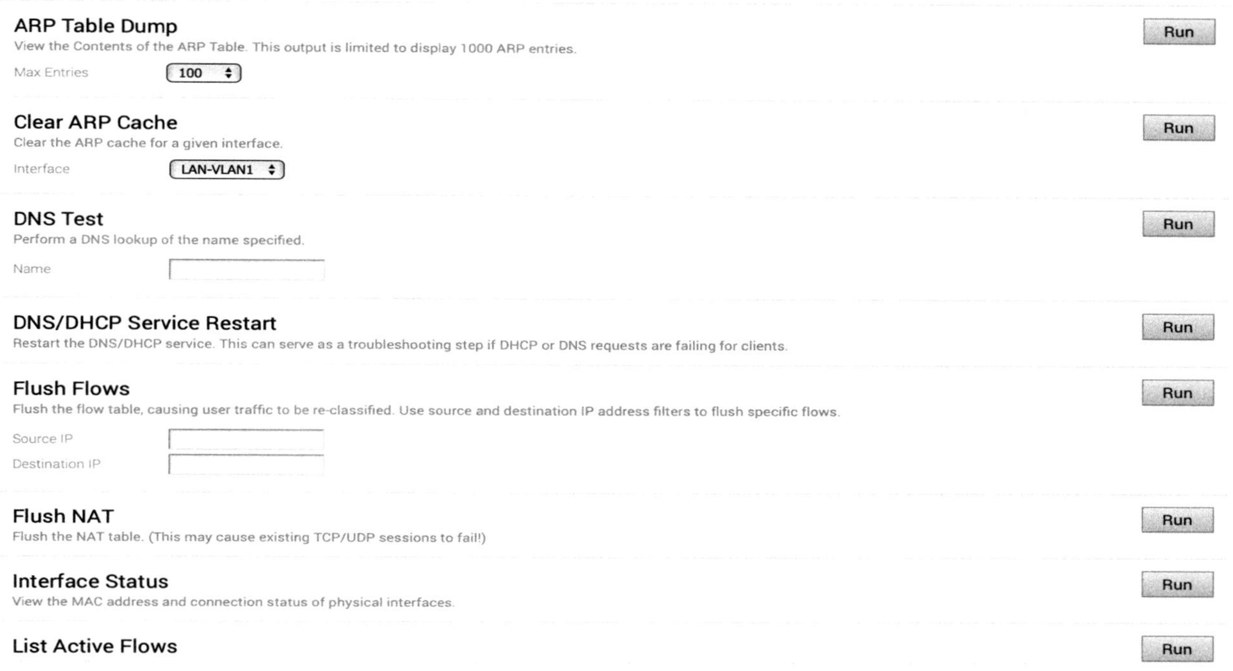

ARP Table Dump
View the Contents of the ARP Table. This output is limited to display 1000 ARP entries.
Max Entries   [100 ♦]   [Run]

Clear ARP Cache
Clear the ARP cache for a given interface.
Interface   [LAN-VLAN1 ♦]   [Run]

DNS Test
Perform a DNS lookup of the name specified.
Name   [   ]   [Run]

DNS/DHCP Service Restart
Restart the DNS/DHCP service. This can serve as a troubleshooting step if DHCP or DNS requests are failing for clients. [Run]

Flush Flows
Flush the flow table, causing user traffic to be re-classified. Use source and destination IP address filters to flush specific flows.   [Run]
Source IP   [   ]
Destination IP [   ]

Flush NAT
Flush the NAT table. (This may cause existing TCP/UDP sessions to fail!)   [Run]

Interface Status
View the MAC address and connection status of physical interfaces.   [Run]

List Active Flows   [Run]

Figure 9.5: Remote Diagnostics

The most commonly used diagnostics are `List Active Flows`, `List Paths`, or `Route Table Dump`. There are also options to do simple tests such as a ping or traceroute. Figure 9.6 shows an example of the `List Paths`, where the current established paths to the primary and secondary gateway of the selected VMware SD-WAN Edge are shown together with tunnel-relevant status such as state, bandwidth, or even jitter.

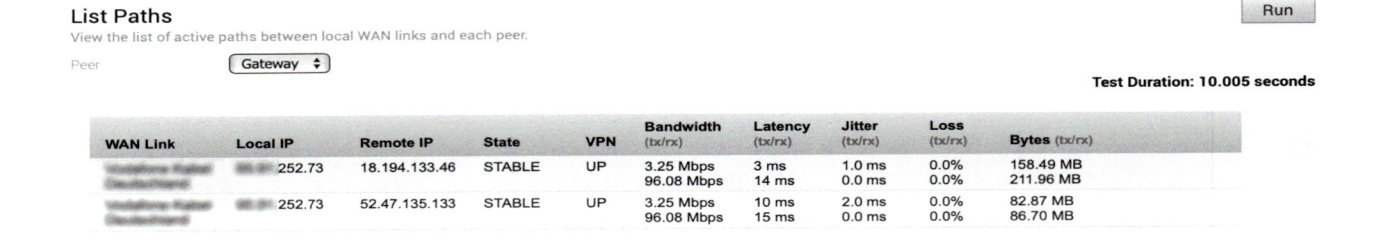

Figure 9.6: List Paths example

Remote Actions

`Remote Actions` are also found in the `Test & Troubleshoot` menu. They are separated from the other diagnostics tools, because most of them have a service impact and require additional confirmation before they are initiated.

There are five options available at this point as seen in Figure 9.7. While all of them are self-explanatory, `Restart Service` is the most commonly used. It just restarts the SD-WAN-specific software components on the VMware SD-WAN Edge in order to make the device re-read the configuration and initiate routing protocols and other services from a clean state. Navigate to `Test & Troubleshoot > Remote Actions`

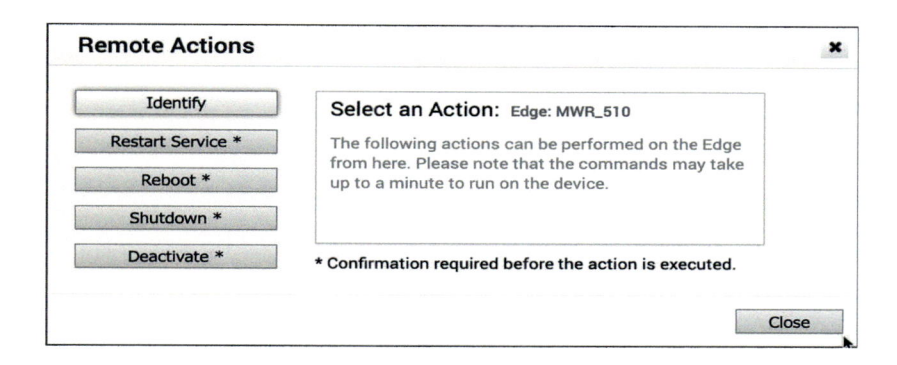

Figure 9.7: Remote Actions

Integration 205

Packet Capture

`Packet Capture` is found in the `Test & Troubleshooting` menu. With the `Request PCAP Bundle` button on the upper right, this allows selection of an interface of the selected VMware SD-WAN Edge and will run a packet capture for a defined time between 5 seconds and 120 seconds. An example is shown in Figure 9.8. Navigate to `Test & Troubleshoot > Packet Capture`

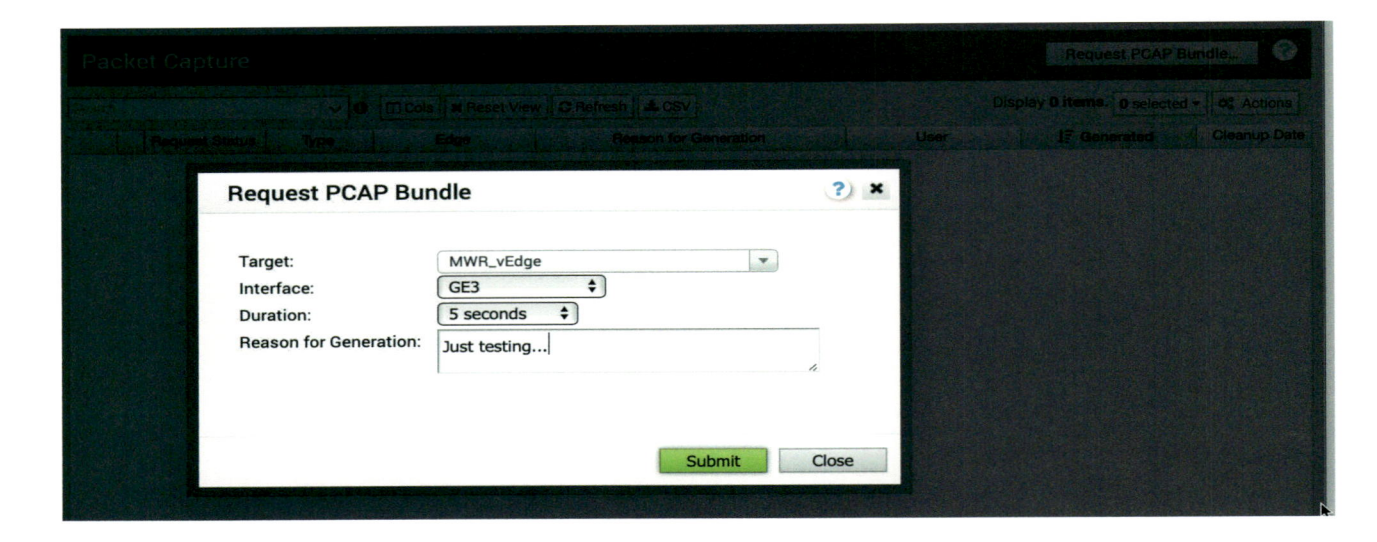

Figure 9.8: Request PCAP bundle

When the capture is shown as complete in the list of capture files, it can be downloaded by clicking on the word **complete**. The zip files contain the PCAP file as well as a JSON manifest showing data about the capture. With the right protocol knowledge and a tool like Wireshark, the capture file can be examined as seen in Figure 9.9.

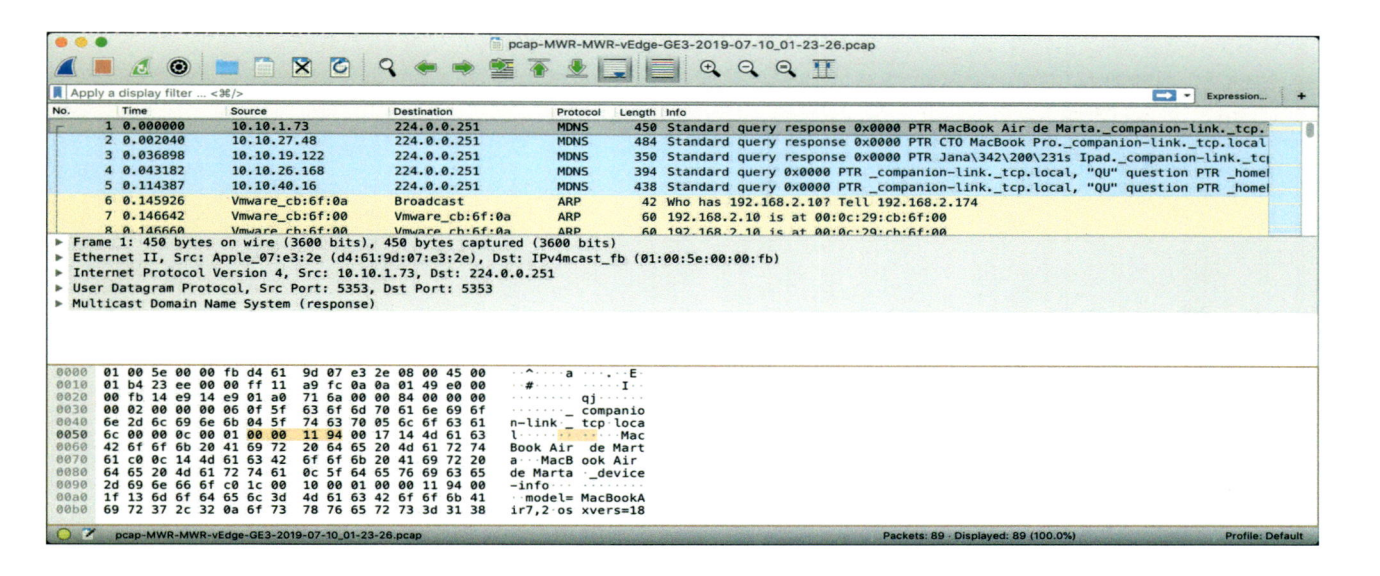

Figure 9.9: Using Wireshark to examine the pcap file

Integration   207

Diagnostic Bundles

When an edge is monitored in the UI of the VMware SD-WAN Orchestrator under administration there is an option listed called **Generate Diagnostic Bundle** which opens the dialogue seen in Figure 9.10.

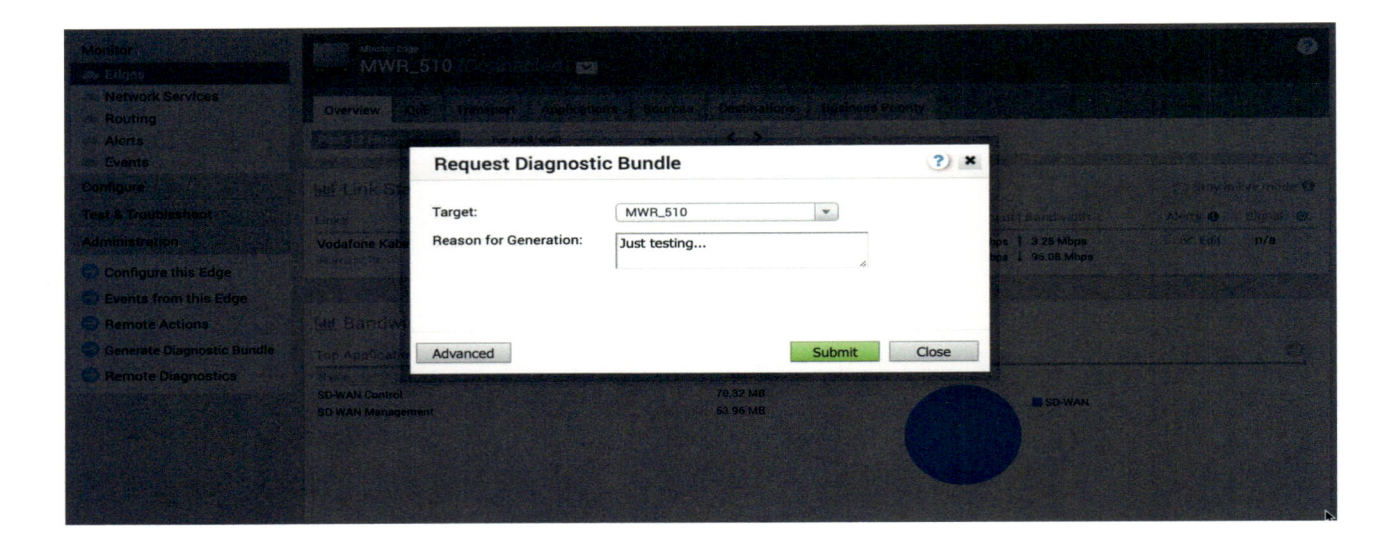

Figure 9.10: Generate diagnostic bundle

This functionality is used when VMware support needs to be included in troubleshooting. Generating a diagnostic bundle collects information from various places on the VMware SD-WAN Edge and packages it up in a ZIP file, which will be stored on the VMware SD-WAN Orchestrator accessible to VMware support.

The collected files in the ZIP archive summarize the state of the VMware SD-WAN Edge at the time of the problem. This includes log files, configuration information, and potential core files. All this helps VMware support or even VMware engineering to narrow down a problem further. Access to diagnostic bundles require operator level access to the VMware SD-WAN Orchestrator to download the files.

The generation of this bundle should happen while the problem exists and should be attached to the service request opened with VMware to work on this problem. It is meant to give VMware support the appropriate information to work with organization's network engineer on a problem resolution.

Troubleshooting Case Studies

This chapter contains some typical problem scenarios experienced in customer networks. The examples show that such problems are readily solvable, and the sample troubleshooting workflows demonstrated may be of value in tackling other issues. These workflows use some of the tools mentioned in the previous part of this chapter and they introduce other areas of the UI to use for diagnosing issues.

Case Study 1—Activation Issues

During activation, the VMware SD-WAN Edge shows an error message on the activation screen after the activation link is clicked. The specific error is `VCO not reachable` as seen in Figure 9.11.

Figure 9.11: Activation Error

While the VMware SD-WAN Edge User Interface during activation allows for checking the configuration or to generate a diagnostic bundle locally, in many cases, the issue can be fixed by checking the following items:

- Change VMware SD-WAN Orchestrator FQDN to IP address as seen in Figure 9.12.

Figure 9.12: FQDN error

- Check `Ignore` next to `Certificate Errors` as seen in Figure 9.13
- Note: This is common when testing with a lab VMware SD-WAN Orchestrator without proper server certificate. It is not recommended to check Ignore Certificate Errors in production environment, as the authenticity of the VMware SD-WAN Orchestrator can not be verified.

Figure 9.13: Certificate Error

- Change the WAN interface order in the config on the VMware SD-WAN Orchestrator such that the public IP is set to the lowest numbered interface. After that, reset the VMware SD-WAN Edge config and generate a new activation link.

- If there is a firewall in the setup, sitting between the VMware SD-WAN Edge and the VMware SD-WAN Orchestrator, check for filters of port 443/tcp and/or the VMware SD-WAN Orchestrator IP address. The firewall log should show whether or not there was a rule violation during the activation attempt.

Following these steps usually resolves all the common issues experienced during activation of a VMware SD-WAN Edge.

Case Study 2 - Overlay Tunnel Issues

After activation, the VMware SD-WAN Edge shows up as connected in the Monitor section of the VMware SD-WAN Orchestrator, but `Connectivity through VeloCloud Service Gateway is down`, as seen in Figure 9.14.

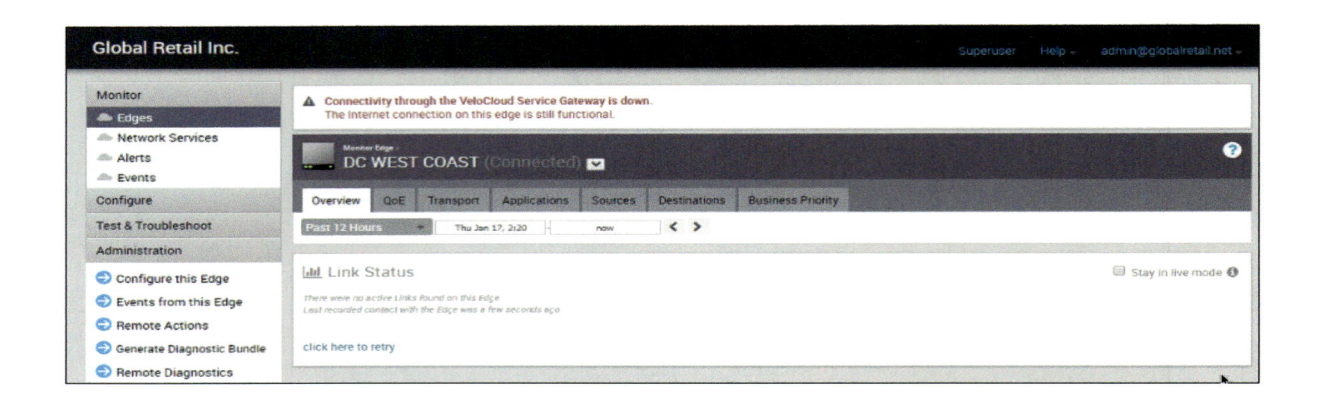

Figure 9.14: Gateway Service unreachable

In many cases the issue can be fixed by checking the following:

- As seen in Figure 9.15, use the aforementioned `List Path` functionality to check, whether there is a path to the `gateway`.

Figure 9.15: No paths

- If there is no path listed, run a `packet capture` to see whether establishing packets for the tunnels are going out (port 2426/udp) as seen in Figure 9.16

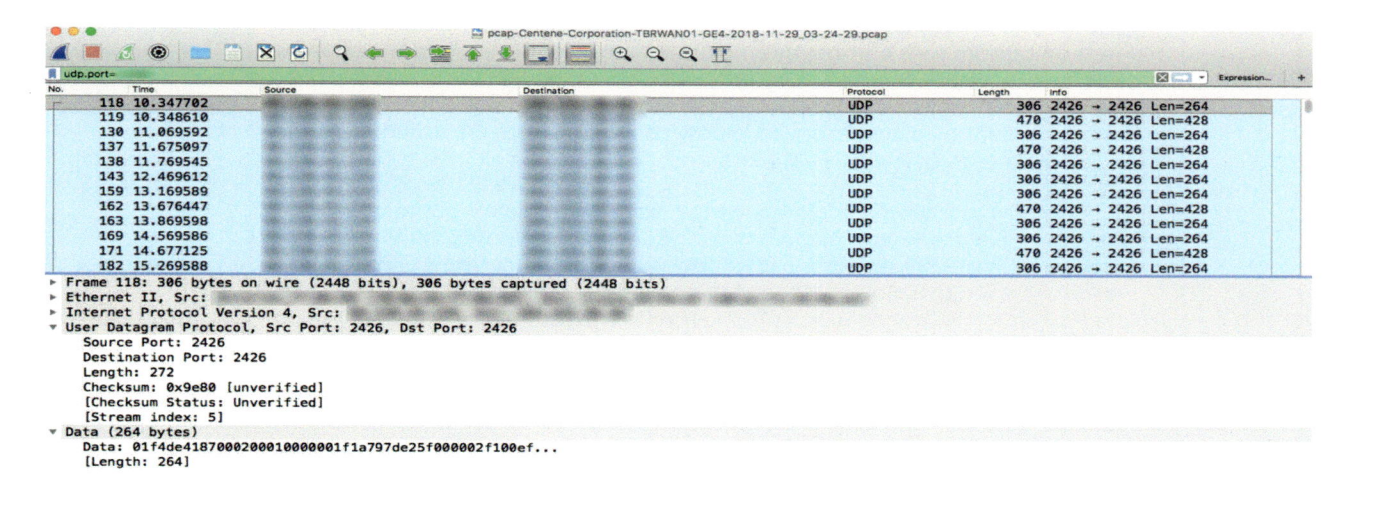

Figure 9.16: No response

- If there are no response packets seen, the issue is almost always a firewall blocking port 2426/udp somewhere between the VMware SD-WAN Edge and the VMware SD-WAN Gateway.

This is the most common problem seen, when trying to get a new SD-WAN network set up.

Case Study 3—Bandwidth Measurement Issues

The VMware SD-WAN solution relies on initial bandwidth measurements for each path in order to make the right decisions for different applications as outlined in Chapter 5: *Application Performance—Dynamic Multi-Path Optimization*. This section shows some of the measures, which can be taken to remediate a situation where the automatic bandwidth measurements show smaller values than what is expected. The measured bandwidth can be seen on the **Overview** page of each VMware SD-WAN Edge in the Monitor section of the VMware SD-WAN Orchestrator as shown in Figure 9.17.

Integration   213

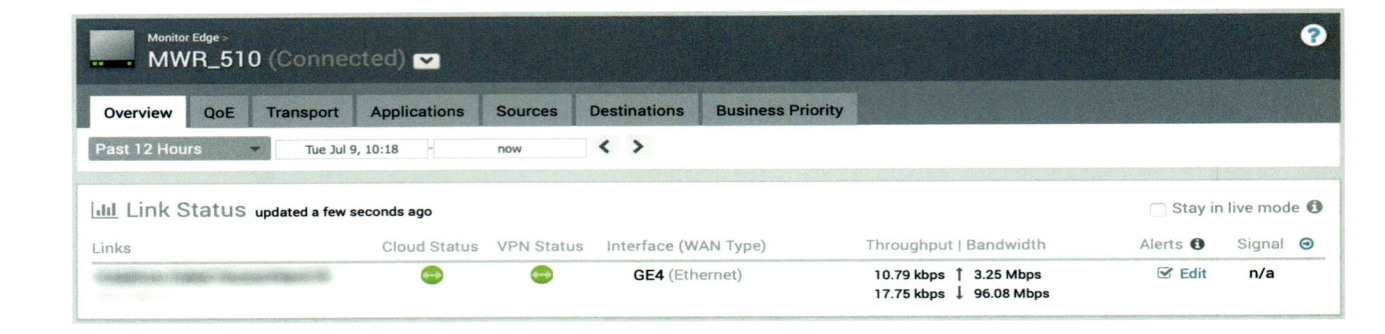

Figure 9.17: Measured bandwidth per interface for a VMware SD-WAN Edge

If these measurements are smaller than expected, the following steps can be done to resolve the problem:

- On the VMware SD-WAN Orchestrator re-initiate a new bandwidth measurement for that interface using the aforementioned `Remote Diagnostics` and the `WAN Link Bandwidth Test` option there as seen in Figure 9.18. Check the measured bandwidth again.

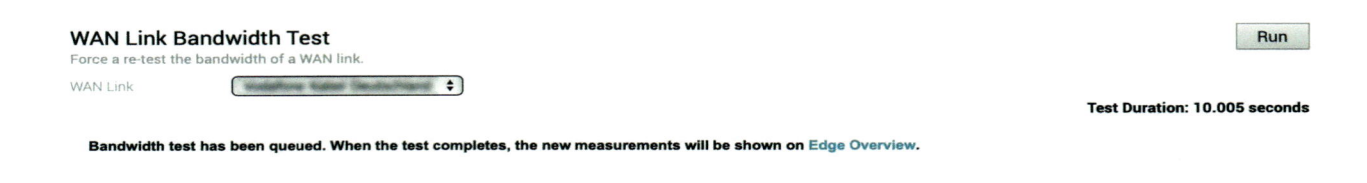

Figure 9.18: Bandwidth remeasurement

- If this didn't resolve the issue, on the VMware SD-WAN Orchestrator go into the configure section for edges and select the VMware SD-WAN Edge, which had the wrong measurements. This should look similar to Figure 9.19. Make adjustments according to Table 9.1. Navigate to `Configure > Edge > {edge name} > Device tab > Edit {overlay_name}`

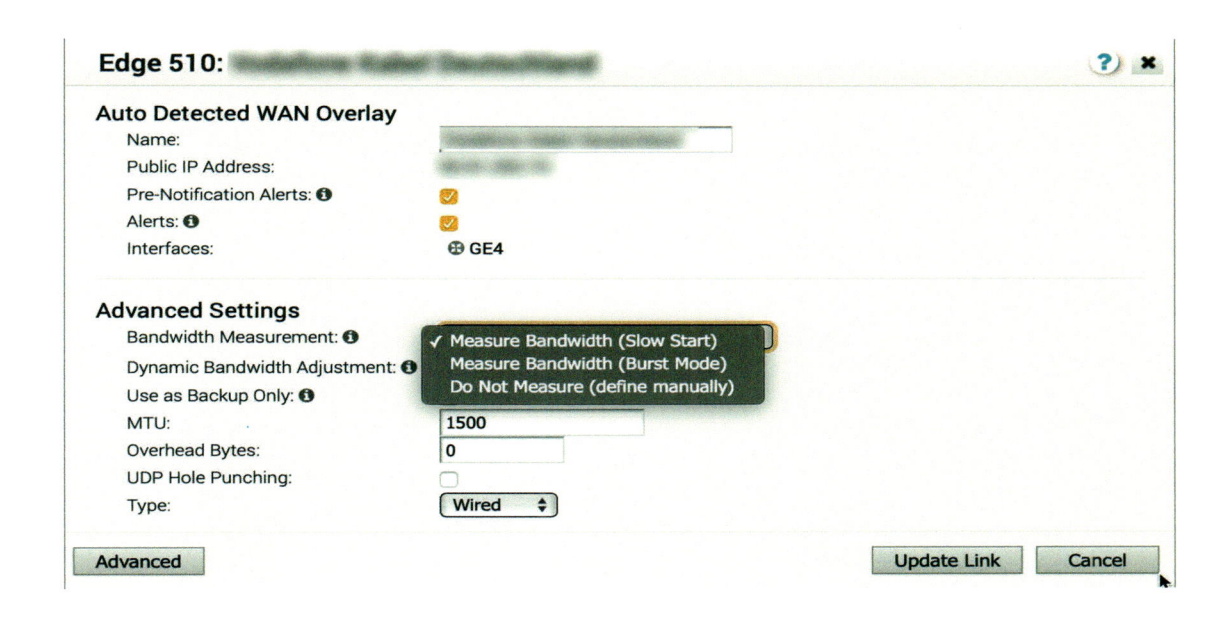

Figure 9.19: Setting the bandwidth settings

Maximum Bandwidth	Recommended Setting
200 Mbps	Slow Start
> 200 Mbps	Burst Mode

Table 9.1: Bandwidth measurement setting recommendations

> **TECH TIP**
>
> In release 3.3.0 and above, if Slow Start measures more than 175 Mbps, the Edge will now automatically switch to the Burst Mode measurement test—which can measure up to the capacity of the Edge.

Integration

Case Study 4—Routing Issues

This chapter should give insight into some of the tools used to identify routing issues. This comes from customers in a site complaining that they cannot reach services at other locations. The following checks can help to find out where the problem could be.

- Rule out any general reachability problems within the infrastructure as described in *Case Study 2* above. Looking at the specific `VMware SD-WAN Edge Overview` page in the `Monitor` section of the VMware SD-WAN Orchestrator gives a good indication of when `VPN Status` and `Cloud status` are green as seen in Figure 9.20.

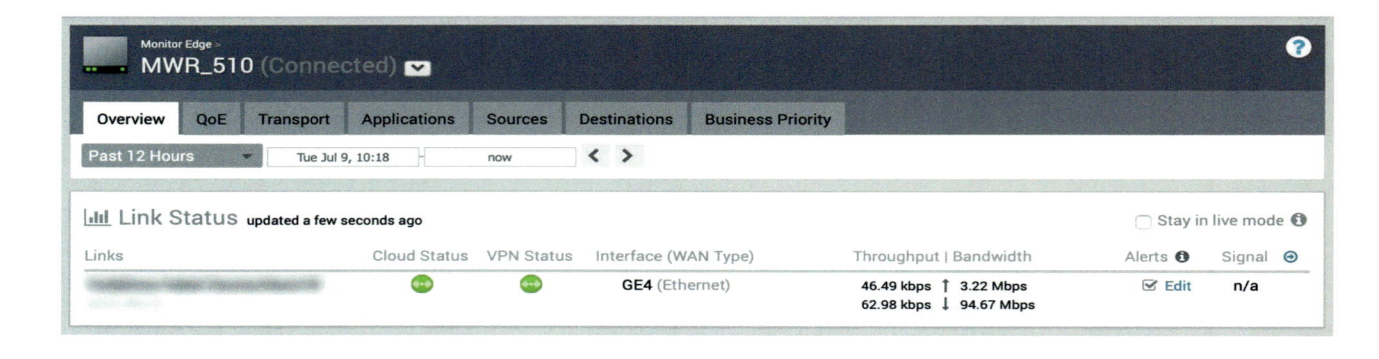

Figure 9.20: Edge Status Overview

- In the `Configure` section of the VMware SD-WAN Orchestrator, check the `Overlay Flow Control` to see whether all routes required for accessibility to the destination the users are attempting to reach are shown for the configured Cloud VPN as seen in Figure 9.21.

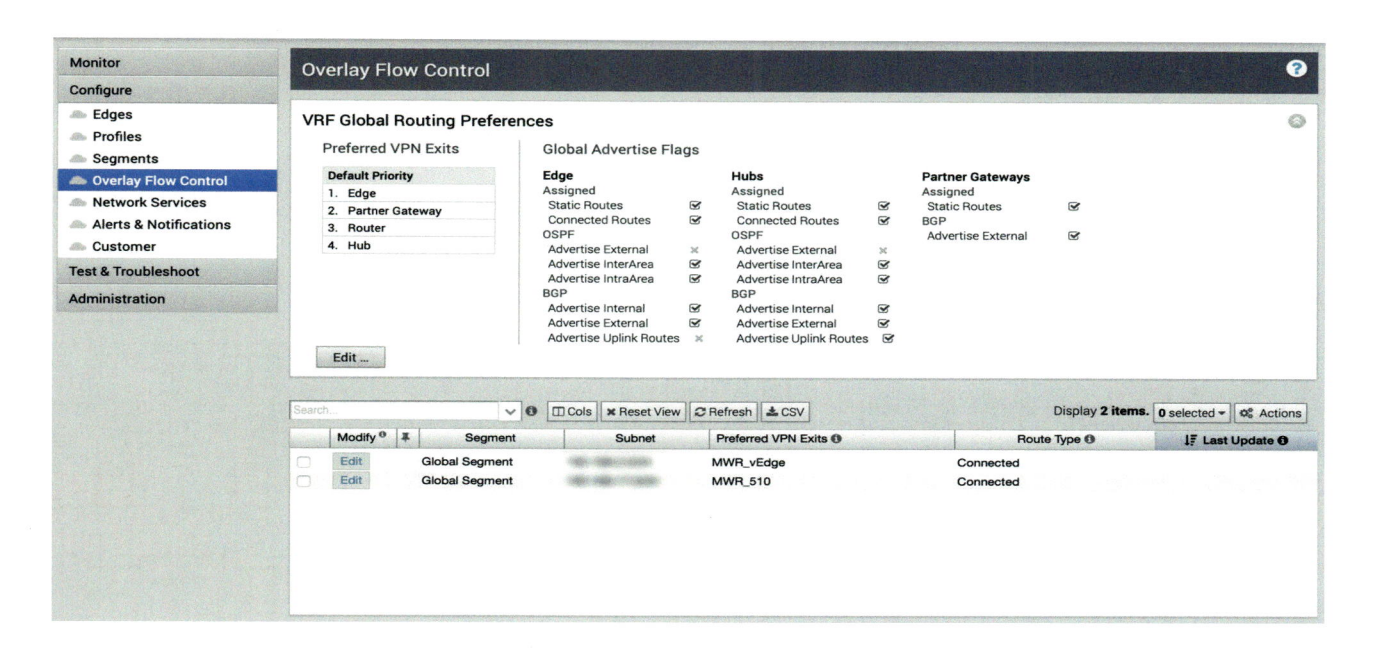

Figure 9.21: Overlay Flow Control

- It is also possible to check the individual routing tables of a VMware SD-WAN Edge using the `Remote Diagnostics` as outlined above. There is an option to dump the SD-WAN overlay routing table locally. A sample can be seen in Figure 9.22.

Integration 217

Route Table Dump
View the contents of the Route Table.

Segment [all ⇳]

Run

Test Duration: 10.005 seconds

Segmented Route Table

Address	Segment	Netmask	Type	Cost	Reachable	Next Hop
	Global	255.255.255.255	Edge	0	TRUE	Cloud VPN
	Global	255.255.255.255	Connected	0	TRUE	br-management
	Global	255.255.255.255	N/A	0	TRUE	GE4
	Global	255.255.255.255	N/A	0	TRUE	
	Global	255.255.255.255	Cloud	0	TRUE	
	Global	255.255.255.0	Connected	0	TRUE	br-network1
	Global	255.255.255.0	Edge	0	TRUE	Cloud VPN
	Global	255.255.255.0	Connected	0	TRUE	GE4
0.0.0.0	Global	0.0.0.0	Cloud	0	TRUE	Cloud Gateway
0.0.0.0	Global	0.0.0.0	Cloud	6	TRUE	GE4

Figure 9.22: Local SD-WAN Overlay Route Table Dump

- `VMware SD-WAN Edge Remote Diagnostics` also has the ability to display dynamically learnt underlay routes (BGP/OSPF) as shown in the Figure 9.23.

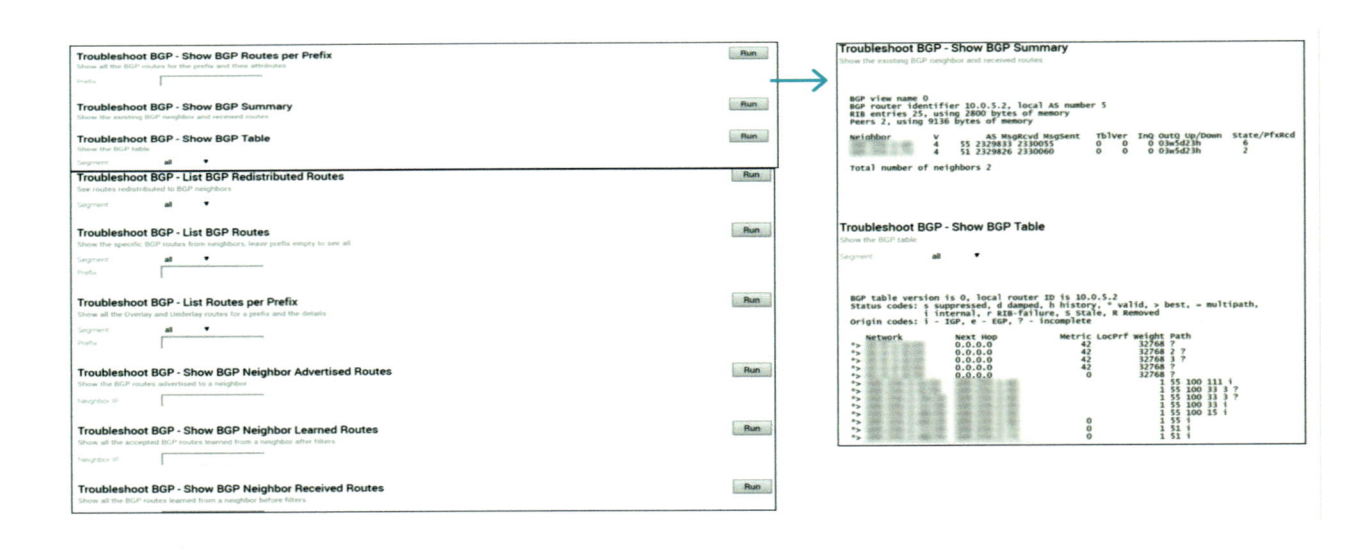

Figure 9.23: Local SD-WAN dynamically learnt underlay routes

218   SD-WAN 1:1

- Finally, the aforementioned **Remote Diagnostics** have **Ping** and **Traceroute** tools available to check whether the VMware SD-WAN Edges can mutually reach themselves.

- If this doesn't fix the problem, revert back to the aforementioned **Packet Capture** in order to get more insight on what the problem can be.

Case Study 5—Application Performance Issues

This case study is mainly used to show some of the areas where application-specific configuration or troubleshooting can happen. The example used is Office365, which seems not to work as expected for customers connected to a specific VMware SD-WAN Edge. Here are some of the steps to quickly identify the issue which can then help to mitigate a problem with a specific application:

- Following the report of the issue, the network engineer navigates to **Test & Troubleshoot > Remote Diagnostics > {Edge name} > List Active Flows command.** The **List Active Flows** output will provide an indication of what flows are going through the Edge, as well as their destination and classification. This output can be further focused on particular sources, destinations, and ports to aid in remediation. A specific source IP of 10.0.213 is used in the example output shown in Figure 9.24.

Src IP	Dst IP	Segment	Protocol	Src Port	Dst Port	Application	Link Policy	Route	Business Policy
10.0.0.213	74.125.142.188	tesaka-lab1	TCP	49946	5228	tcp	Loadbalance	Direct to Cloud	Default-Internet-Other
10.0.0.213	172.217.164.106	tesaka-lab1	TCP	57527	443	google_gen	Loadbalance	Direct to Cloud	Web
10.0.0.213	52.114.74.43	tesaka-lab1	TCP	57529	443	office365	Loadbalance	Direct to Cloud	Office365 MultiPath
10.0.0.213	172.217.6.78	tesaka-lab1	TCP	57494	443	google_gen	Loadbalance	Direct to Cloud	Web
10.0.0.213	52.37.243.173	tesaka-lab1	TCP	55538	443	https	Loadbalance	Direct to Cloud	Web
10.0.0.213	52.114.74.43	tesaka-lab1	TCP	57525	443	office365	Loadbalance	Direct to Cloud	Office365 MultiPath
10.0.0.213	216.58.195.74	tesaka-lab1	TCP	56963	443	https	Loadbalance	Direct to Cloud	Web
10.0.0.213	13.107.6.171	tesaka-lab1	TCP	50004	443	office365	Loadbalance	Direct to Cloud	Office365 MultiPath
10.0.0.213	13.107.6.171	tesaka-lab1	TCP	57530	443	office365	Loadbalance	Direct to Cloud	Office365 MultiPath

Figure 9.24: List Active Flows showing Office365 going Direct

- From the output in Figure 9.24 it can be seen that the traffic is matching the correct business policy of Office365 Multi-Path however the policy has been incorrectly configured to send this critical traffic **Direct** instead of **Multi-Path**. In order to change the business policy navigate to **Configure > Profile > {Profile Name} > Business Policy tab > Edit {Business Policy Name}.** Figure 9.25 shows the incorrect configuration and Figure 9.26 shows the corrected configuration.

Integration

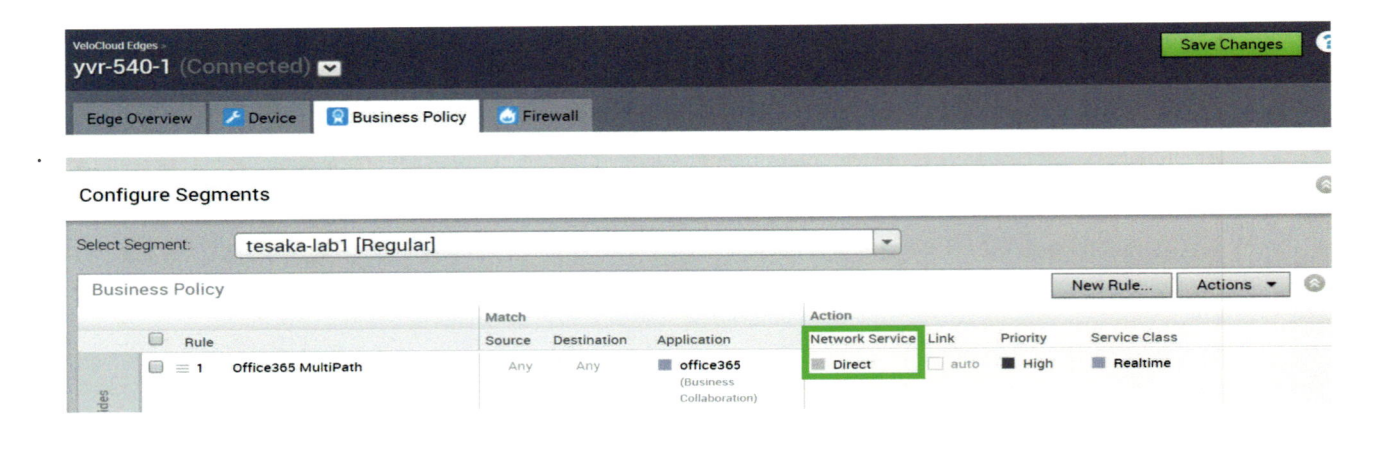

Figure 9.25: Incorrect Business Policy for Office365

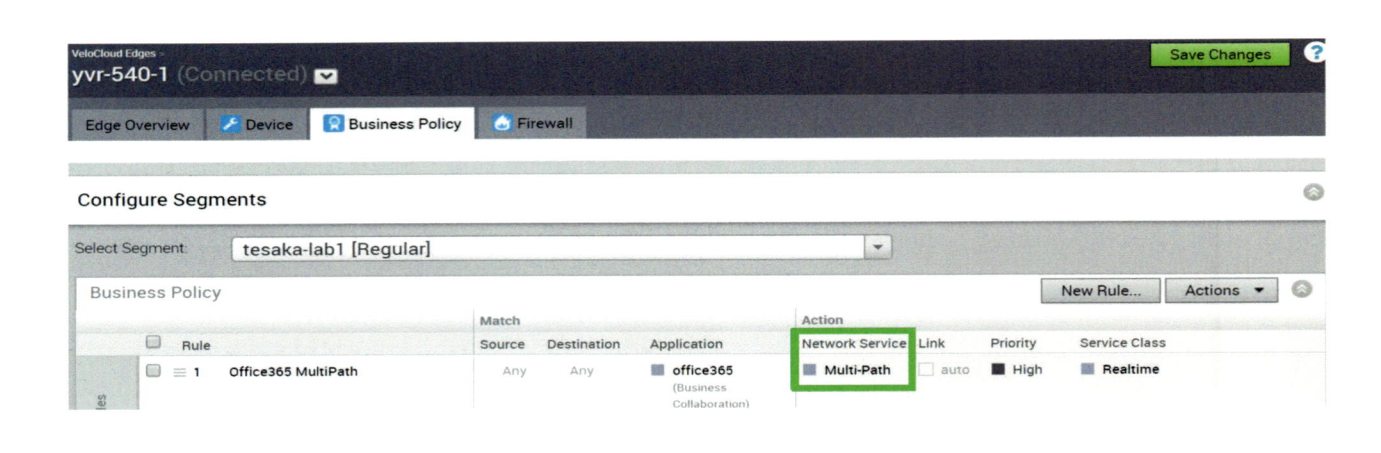

Figure 9.26: Correct Business Policy for Office365

After the entry for Office365 was fixed to use `Multi-Path` instead of `Direct` for the `Network Service`, the issue for the users was resolved. Further evidence of this can be seen by looking at the `Active Flows` again (Figure 9.27).

10.0.0.213	52.37.243.173	tesaka-lab1	TCP	55538	443	https	Loadbalance	Direct to Cloud	Web
10.0.0.213	52.114.77.34	tesaka-lab1	TCP	57560	443	office365	Replicate	Cloud via Gateway	Office365 MultiPath
10.0.0.213	52.114.128.4	tesaka-lab1	TCP	57559	443	office365	Replicate	Cloud via Gateway	Office365 MultiPath
10.0.0.213	13.107.6.171	tesaka-lab1	TCP	57530	443	office365	Replicate	Cloud via Gateway	Office365 MultiPath
10.0.0.213	172.217.164.106	tesaka-lab1	TCP	57526	443	google-gen	Loadbalance	Direct to Cloud	Web

Figure 9.27: List Active Flows now showing Office365 being sent via the gateway

---

Key Takeaways:

1. VMware SD-WAN Orchestrator UI has a variety of tools for troubleshooting.
2. The tools available range from basic network tools such as ping or traceroute to very sophisticated displays of link status and remediation results.
3. VMware SD-WAN Orchestrator UI List Active Flows from Remote Diagnostics provides an indication of what flows are going through the VMware SD-WAN Edge device.
4. VMware SD-WAN Orchestrator UI also holds all the configuration for the SD-WAN Overlay, necessary changes can be initiated from the same session.
5. VMware SD-WAN Orchestrator UI **Remote Diagnostics** tab has a collection of commands used to display various states of information available in the VMware SD-WAN Edge.
6. There are also tools available to help the VMware support to narrow down issues, which require fixes to the software itself by collecting all required information during a problem state with a single click.

Integration

vRealize Network Insight for SD-WAN

VMware has recently enhanced one of its most popular network monitoring tools to include support for the VMware SD-WAN solution. The tool is called **vRealize Network Insight (vRNI)** and is an optional tool which provides additional monitoring and troubleshooting capabilities beyond those which are built into the VMware SD-WAN solution. The vRNI tool collects data from network and security elements and IPFIX data to analyze the topology and flows throughout physical and overlaying virtual networks and endpoints. As seen in Figure 9.28 the tool then renders highly intuitive graphical depictions of end-to-end traffic patterns, network utilization, network alerts, and even some high-level Return-On-Investments (ROI) predictions for SD-WAN modeling.

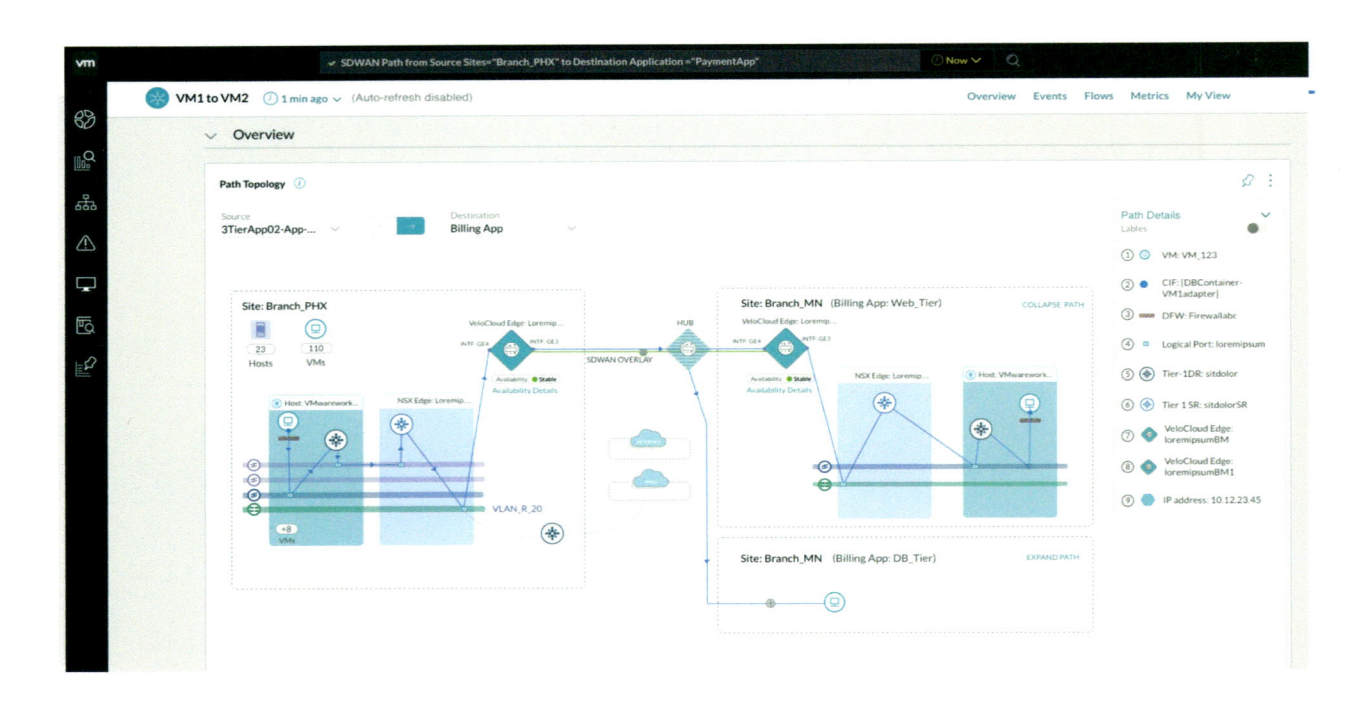

Figure 9.28: vRealize Network Insight sample output

222   SD-WAN 1:1

API

The VMware SD-WAN Orchestrator provides REST, Java, and Python interfaces that enable integration with third-party systems. The APIs can be leveraged externally for tasks such as configuration, management, custom dashboard for monitoring and troubleshooting, and OSS/BSS integration. This chapter discusses how a network engineer can leverage the REST API for VMware SD-WAN. It is meant as an introduction as a full programming manual for the API is beyond the scope of this book. Relevant links for further readings are provided.

The VMware SD-WAN Orchestrator REST API is a wrapper around the JSON-RPC API and provides an easy-to-use interface for developers familiar with REST semantics for request and response. Due the underlying JSON-RPC implementation, the following considerations apply when interfacing with the VMware SD-WAN Orchestrator using REST APIs.

- All the REST APIs are HTTP Post only. Request data must be passed as JSON encoded payload in the HTTP request body.

- Access to the REST APIs is achieved after successful authentication and receiving a session cookie via the `login/enterpriseLogin` or `login/operatorLogin` methods.

    - Enterprise and partner (MSP) users: `login/enterpriseLogin`
    - Operator users: `login/operatorLogin`

- VMware SD-WAN Orchestrator login information is stored as session cookie `velocloud.session`. This information is used as authentication token for subsequent API calls to the VMware SD-WAN Orchestrator.

Interfacing with VMware SD-WAN Orchestrator with REST API

> **TECH TIP**
>
> The VMware SD-WAN Orchestrator user interface is built on top of the provided REST API itself

There are many ways to interface the VMware SD-WAN Orchestrator with REST API. Chrome DevTools works with APIs as it is built-into the browser and offers the ability to observe the transactions between the web client and the VMware SD-WAN Orchestrator on the API exchanged. In Figure 9.29 there is an example where the network engineer created a new branch. As they are clicking through and filling out the branch information (name, model, profile etc.) using the VMware SD-WAN Orchestrator portal, the format and the API calls are being captured by Chrome DevTools.

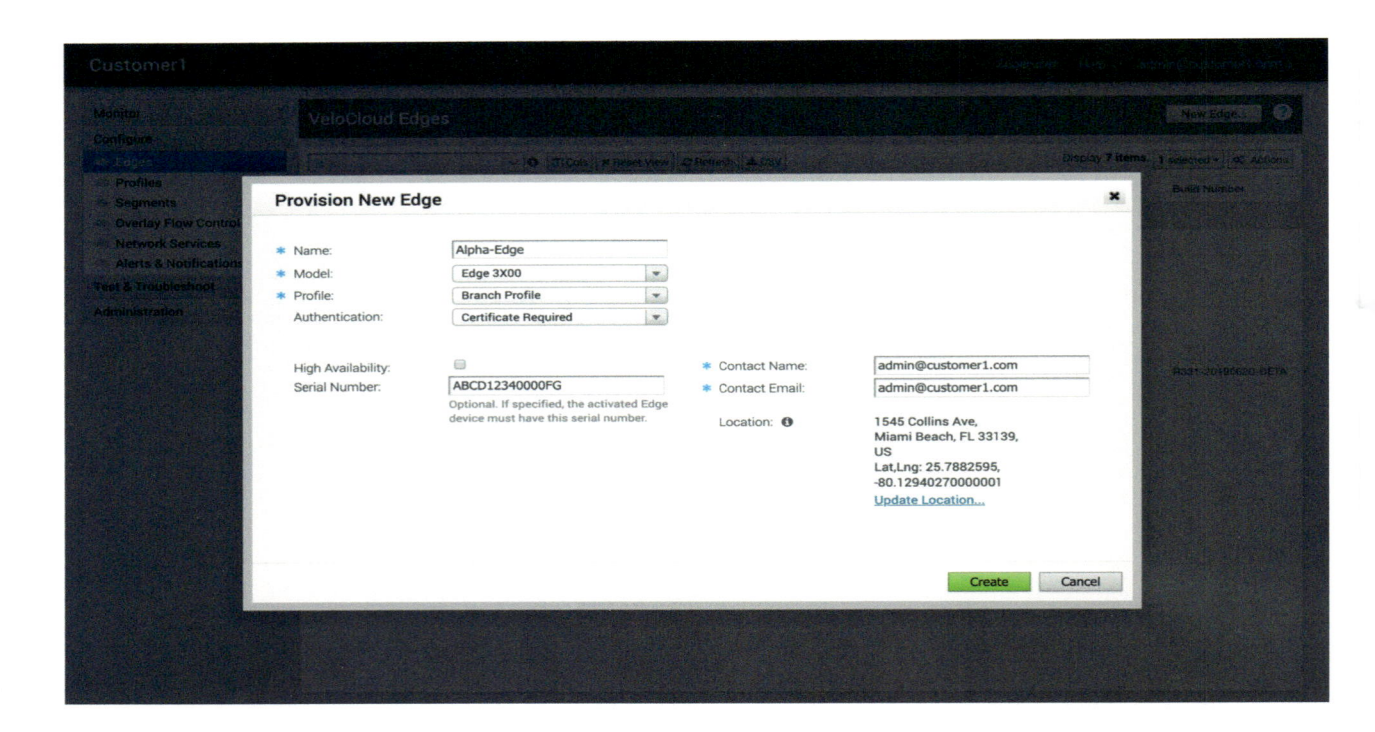

Figure 9.29: Create VMware SD-WAN Edge form

The network engineer can then open up the DevTools and examine the REST API calls and JSON-RPC contents exchanged. Figure 9.30 shows an example of output from Chrome DevTools for a SD-WAN Edge creation procedure. It contains the headers, encoding format which is application/json, and the JSON-RPC payload with the fields necessary to populate during a branch creation process. Using Chrome DevTools is a good way for a network engineer to become familiar with the REST API structures of the SD-WAN solution.

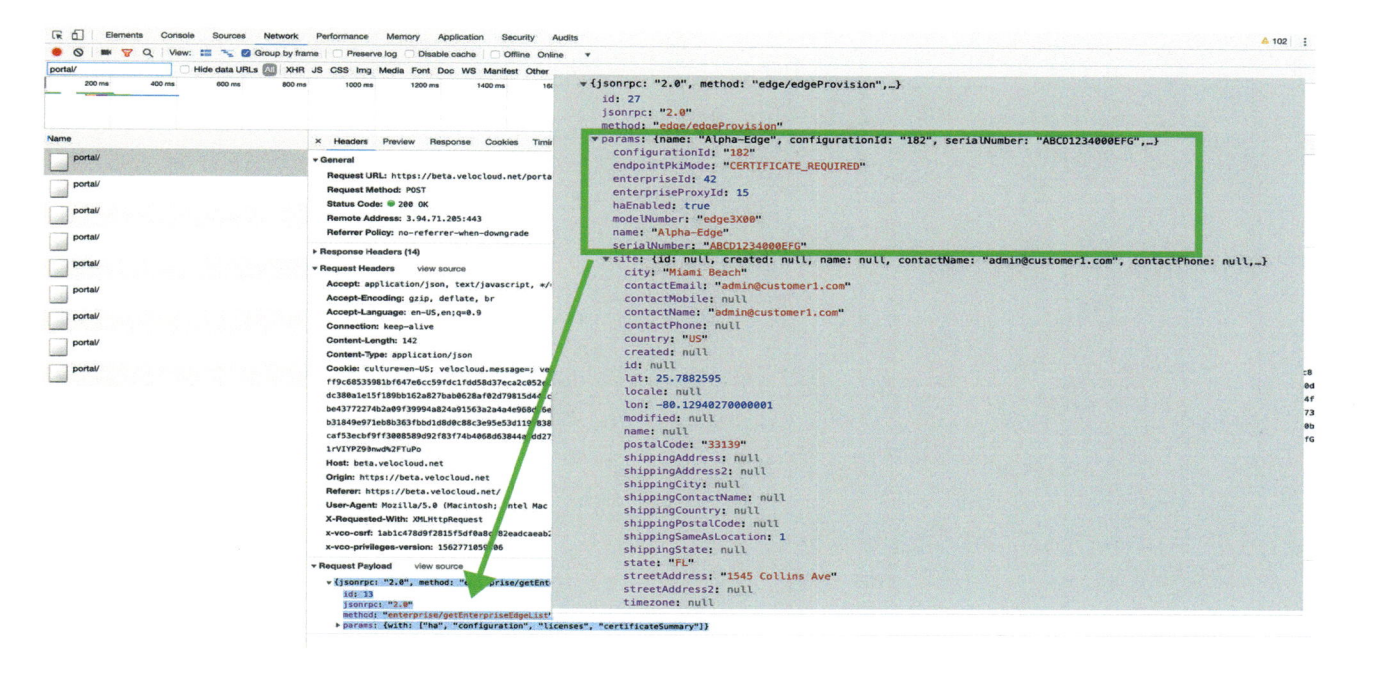

Figure 9.30: Chrome DevTools Output for creation of a new VMware SD-WAN Edge

Postman

Postman is another popular tool for REST API workflow and interactions. Using Postman, a network engineer can create a new edge similar to using the web portal of the VMware SD-WAN Orchestrator. To use REST API, there are a few basic structures for making the calls.

The following URL will be the base for the API call:

```
https://<orchestrator_hostname>/portal/rest/
```

Followed by the operation the network engineer is looking to perform, e.g.

```
/edge/edgeProvision
```

Is for provisioning an Edge device. The complete URL for the REST API will look like

```
https://<orchestrator_hostname>/portal/rest/edge/edgeProvision
```

Integration   225

For a complete set of REST API repository supported by VMware SD-WAN, visit the site below

```
https://code.vmware.com/apis/671/velocloud-sdwan-vco-api
```

Figure 9.31 is an example of using Postman to create a new VMware SD-WAN Edge.

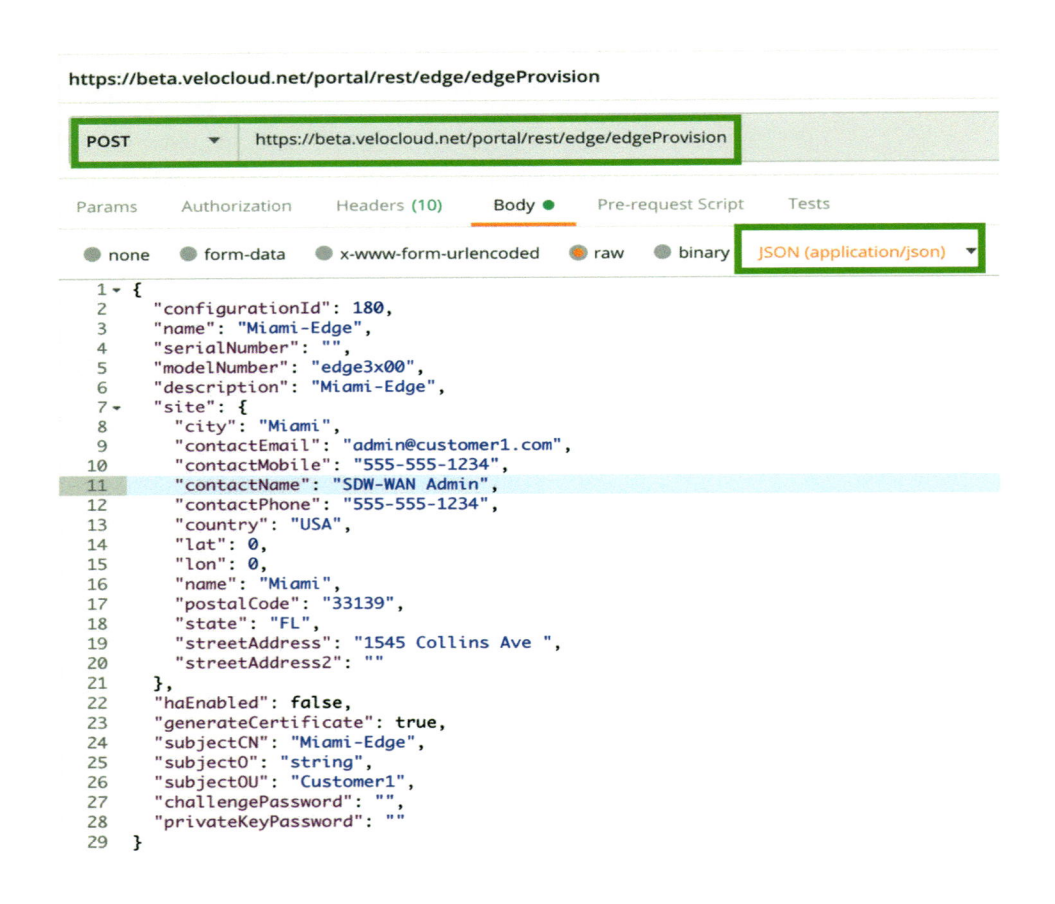

```
https://beta.velocloud.net/portal/rest/edge/edgeProvision

POST    ▼    https://beta.velocloud.net/portal/rest/edge/edgeProvision

Params    Authorization    Headers (10)    Body ●    Pre-request Script    Tests

● none    ● form-data    ● x-www-form-urlencoded    ● raw    ● binary    JSON (application/json) ▼

 1 ▾ {
 2      "configurationId": 180,
 3      "name": "Miami-Edge",
 4      "serialNumber": "",
 5      "modelNumber": "edge3x00",
 6      "description": "Miami-Edge",
 7 ▾    "site": {
 8          "city": "Miami",
 9          "contactEmail": "admin@customer1.com",
10          "contactMobile": "555-555-1234",
11          "contactName": "SDW-WAN Admin",
12          "contactPhone": "555-555-1234",
13          "country": "USA",
14          "lat": 0,
15          "lon": 0,
16          "name": "Miami",
17          "postalCode": "33139",
18          "state": "FL",
19          "streetAddress": "1545 Collins Ave ",
20          "streetAddress2": ""
21      },
22      "haEnabled": false,
23      "generateCertificate": true,
24      "subjectCN": "Miami-Edge",
25      "subjectO": "string",
26      "subjectOU": "Customer1",
27      "challengePassword": "",
28      "privateKeyPassword": ""
29  }
```

Figure 9.31: Usage of Postman to generate a new VMware SD-WAN Edge

> **TECH TIP**
>
> Notice the REST API is using HTTP POST with application/json as content-type.

Python Scripts

While many tasks found within the UI of the VMware SD-WAN Orchestrator are simple and direct to accomplish, to effectively reach larger scales for certain tasks, it becomes imperative to leverage the API for the purposes of automation. As with any programming language, there is an assumption in this chapter of at least some programming knowledge to make effective use of the API. One such example would be to retrieve a list (output snipped) of edges configured within a particular enterprise as demonstrated here:

Request

```
{
        "id": 14,
        "jsonrpc": "2.0",
        "method": "enterprise/getEnterpriseEdges",
        "params": {
            "enterpriseId": 1,
            "with": [
                "site"
            ]
        }
    }
```

Response

```
{
   "jsonrpc": "2.0",
   "result": [
     {
       "id": 1,
       "created": "2017-01-22T01:47:36.000Z",
       "edgeHardwareId": null,
       "enterpriseId": 1,
       "siteId": 3,
       "activationKey": "7M6W-QQYH-V369-SC4W",
       "activationKeyExpires": "2017-02-21T01:47:36.000Z",
       "activationState": "ACTIVATED",
       "activationTime": "2017-01-22T01:57:02.000Z",
       "softwareVersion": "2.4.1",
       "buildNumber": "R241-20170507-QA",
       "factorySoftwareVersion": "3.0.0",
       "factoryBuildNumber": "R30-20170119-BETA",
       "softwareUpdated": "2017-05-07T22:17:42.000Z",
       "selfMacAddress": "00:50:56:9c:44:94",
       "deviceId": "00:50:56:9c:44:94",
```

The above call references some additional information that would need to be determined in order to get the correct results. The first and most important is to use VMware SD-WAN Orchestrator login information to be stored as a cookie with the orchestrator authentication token for subsequent calls to be made. The second would be to derive the enterprise ID if logged in as an operator or MSP. This ID can be derived from the URL in the browser, as seen in Figure 9.32:

Figure 9.32: Enterprise ID

There are client libraries that are code-generated, however, the VMware-recommended client for consumption of the API can be found here:

https://code.vmware.com/samples/5554/velocloud-orchestrator-json-rpc-api-client---python

In addition, there are example calls to accomplish various tasks that can be found at the following link. This list is public domain and will continue to be updated:

https://code.vmware.com/apis/671/velocloud-sdwan-vco-api

> Key Takeaways:
>
> 1. The VMware SD-WAN solution provides REST APIs on the VMware SD-WAN Orchestrator.
> 2. All APIs are HTTP POST only and require to pass request data as JSON encoded payload.
> 3. Documentation, examples, and clients for API consumption are publicly available through https://code.vmware.com.

Alvina is encouraged by what she has learned about VMware SD-WAN and from the nods and affirmations in the room from the NOC team and Rodney were also hopeful. Rodney and the NOC team learned that managing the VMware SD-WAN solution is not overly difficult, but it is a bit of a change to focus on monitoring of the VMware SD-WAN Orchestrator instead of focusing on the VMware SD-WAN Edges as would be more typical in their traditional WAN.

The demo of the VMware vRealize Network Insight (VRNI) tool was particularly enlightening. VRNI supports the ability to generate an assessment and ROI and those impressive graphics would impress the Board! Rodney had a 'lightbulb' moment when he saw that VRNI could graphically display an end-to-end path through the network. The truly enlightening aspect was the mapping between virtual and physical elements in Allofu's network. The gap between physical and virtual elements has always been a pain point to networking and security teams. Given that the VRNI tool can also provide the same visibility into the data center, both Rodney and the NOC team were excited to demo it right away. Hans actually made a grunting sound they all took to mean he approved, didn't hate it, and frankly, nobody is certain that Hans even likes Hans, so this seems, like a ringing endorsement.

Alvina consults with her team and elects to move forward with the hosted solution from Circuits 'R' Us. Her decision was largely based on the fact that Circuits 'R' Us is building out their own private cloud and Alvina is feeling bullish about some of the new services they will offer. The managed service also takes some of the day-to-day burden off of her team, which helps free up Rodney in particular for more strategic projects. Rodney has already visibly lost weight, which Alvina assumes is directly linked to the decrease in Rodney's T&E expenses for the quarter as a result of far fewer traveling steak dinners. Yves is now teaching goat yoga, but the music has gotten decidedly edgier. The migration to VMware SD-WAN is well underway and has gone very smoothly thus far, though some awkward conversations have occasionally arisen in regard to what Allofu's users consider 'business critical'. Matty reached out to Alvina to make a case for Snapchat, but this was quickly dismissed. The Board of Directors was thoroughly impressed with what Alvina and her team have accomplished and the returns for Allofu. In the words of the Big Lebowski, "It really tied the room together." - Great work Alvina!

10. SD-WAN in the Bigger VMware Picture

SD-WAN in The Virtual Cloud Network

The virtual cloud network (VCN) is VMware's architectural vision for the evolution of the Cloud and the internetworking which provides the underlying virtualized connectivity. A virtual cloud network is the next big evolutionary step of the network, and it is optimized for cloud-first organizations. A VCN is a single network abstraction that creates operational consistency, regardless of the underlying hardware or services and connects everything from the branch to the cloud to the data center. It provides pervasive connectivity for users to apps and for businesses to data, as well as intrinsic security, regardless of location. A VCN realizes the vision of connecting any user to any application in any location. VMware SD-WAN in an integral part of the virtual cloud network vision, as it provides the connectivity layer across different transports and administrative domains.

With a VCN, the branch, public cloud, SaaS, user edge, IoT edge, and data center all run on a single, common network with a consistent set of services (as shown in Figure 10.1).

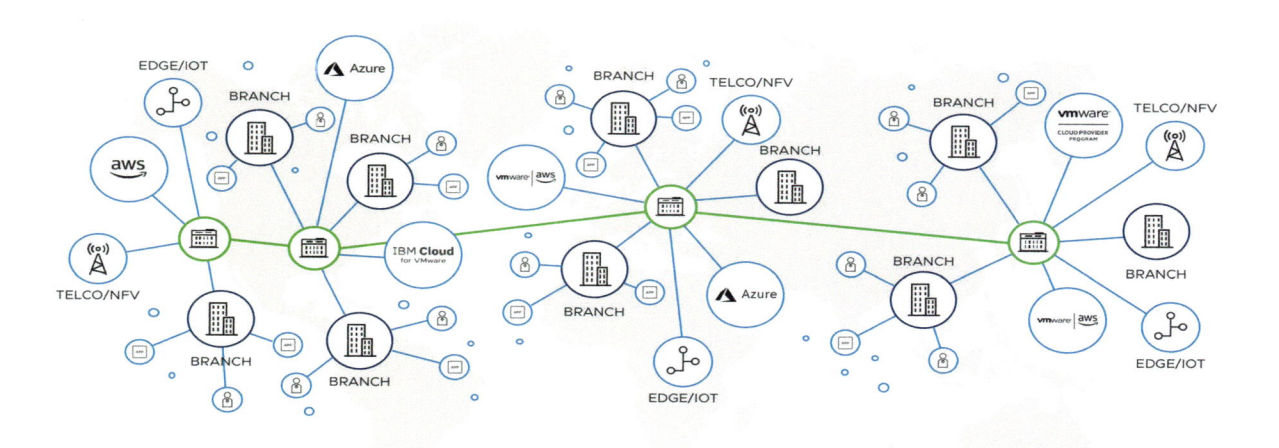

Figure 10.1: Virtual Cloud Network

This is possible because a VCN is built on the same principles that underlie the cloud. Consequently, a VCN can equally meet the needs of everything from the smallest companies to the largest cloud-scale organizations.

The key characteristics of a VCN are as follows:

- **Automated**: The provisioning, deployment, and management of network and security services across the enterprise have historically been done manually, leading to high amounts of human error and slow deployment cycles. Automation can significantly speed this up, reduce and eliminate unplanned downtime. Also, automation will free up valuable engineering time, enabling those resources to focus on innovation and strategic initiatives.

- **Programmable**: Application development and network operations are coming together, and programmers need the ability to access network services in order to build differentiated applications that create competitive advantages.

- **API-accessible**: The days of configuring the network and accessing information through a command line interface (CLI) have come to an end.

SD-WAN in the Bigger VMware Picture

Obviously, application developers will prefer to interface with the network via application programming interfaces (APIs), but so should network engineers who can manage the network through the same interfaces. This can greatly reduce the complexity associated with network operations.

- **Elastic**: The cloud enables organizations to add compute and application resources on demand and then scale up capacity quickly. The network requires the same level of elasticity to ensure applications are performing optimally, regardless of where workers, applications, and data are located.

- **End-to-end**: Network operations teams have historically treated the network as a set of discrete places, such as the campus, branch, data center, and edge. Applications, of course, traverse all of these and require consistency in the areas of performance and security. A virtual cloud network considers the network in its entirety and removes these silos.

- **Autonomous**: Over time, a virtual cloud network will be autonomous and dynamically reconfigured as the business environment changes in response to business policy. For example, if an organization decides that it's safest to put all IoT devices in a single, secure zone, and then something is moved out of the zone, the network should adapt to extend that segment.

- **Machine learning based:** The network generates massive amounts of network data—far too much for even the most experienced engineer to connect the dots and understand what is happening. Machine learning can be used to recognize breaches, spot congestion points that might be impairing application performance, and even predict when issues will occur, so corrective action can be taken before these issues impact the business.

- **Application-centric:** Traditional networks are designed without considering what types of applications are currently running on them and what kinds of services and resources are required. A virtual cloud network is built with the needs of all types of applications—both cloud and on-premises—in mind.

- **Consistent operations from the data center to branch:** With a VCN, networks will have operational consistency in all parts of the network. This will significantly reduce the requirement to have high-level engineers dedicate huge chunks of time performing repetitive tasks.

It's important to understand that with a VCN, architecture matters. Without the necessary building blocks that utilize software-defined networking (SDN) principles (which are virtual, flexible, scalable, and cloud-based) a true VCN can't happen. VCNs must have all these characteristics for a seamless, integrated model to work. Without it, the entire structure falls apart.

It's unlikely that any organization, no matter how technically astute, will be able to migrate its entire network to a VCN overnight. This raises the question of where the shift to a VCN should begin. For many businesses, a software-defined WAN is the best starting point.

VMware SD-WAN is an integral part of the set of solutions that constitute the various components of the virtual cloud network, as shown in Figure 10.2. The individual components are as follows:

- VMware NSX Data Center, which provides networking and security for all workloads inside the data center.
- VMware NSX Cloud, which provides cloud-native network services inside the public cloud.
- VMware NSX Hybrid Connect, which provides data center and cloud workload migration at scale.
- **VMware SD-WAN by VeloCloud, which provides WAN connectivity services between branches, between branches and data centers, and onramp to cloud-based services.**

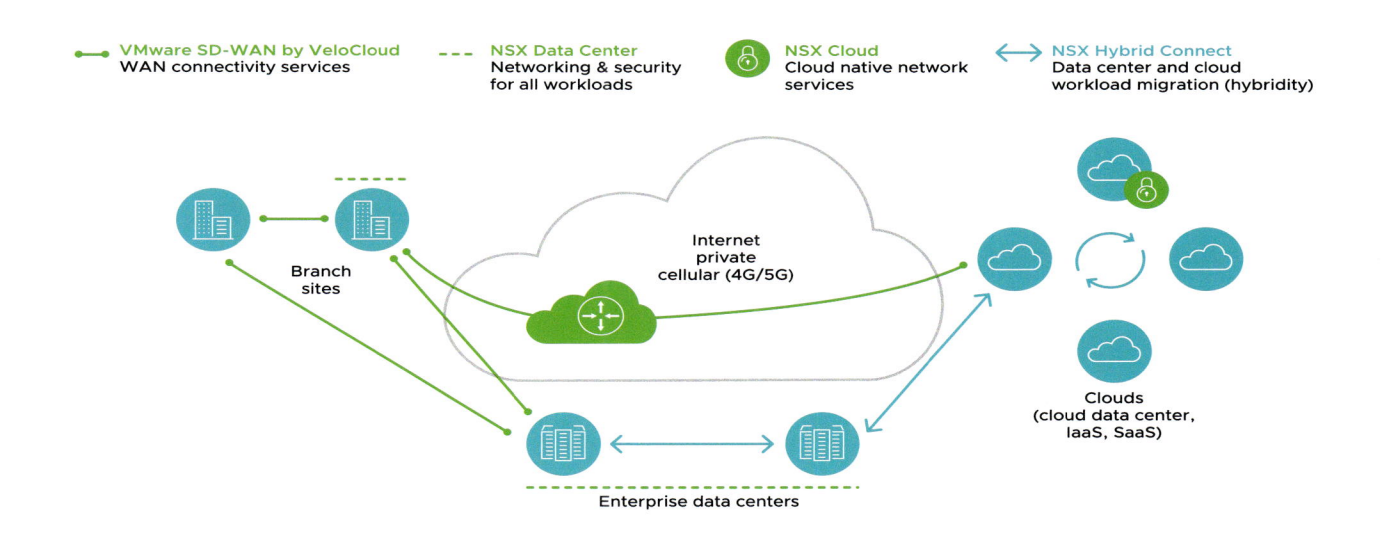

Figure 10.2: Virtual Cloud Network communication elements

SD-WAN in the Bigger VMware Picture

235

VMware SD-WAN and vCloud Network Function Virtualization

VMware SD-WAN as a virtual network function

This chapter shows how VMware SD-WAN can be implemented as a set of Virtual Network Functions (VNFs) on top of a vCloud NFV solution, via a set of scenarios. This chapter is not meant to be a primer on *vCloud NFV*, VMware's Network Function Virtualization Infrastructure (NFVI) solution.

The vCloud NFV platform provides a comprehensive, service-oriented solution, leveraging a cloud computing model that allows ubiquitous, programmatic, on-demand access to a shared pool of compute, network, and storage resources. The vCloud NFV platform is based on standards developed by the [European Telecommunications Standards Institute](#) (ETSI). The solution is integrated with holistic operations management and service assurance capabilities, empowering the communication service provider to rapidly deliver services while ensuring their quality. The vCloud NFV infrastructure delivers myriad telecommunications use cases and facilitates reusability of the service catalog of VNFs. The vCloud NFV platform delivers a complete, integrated solution that has been rigorously tested to ensure compatibility, robustness, and functionality. Components used in creating the solution are currently deployed across many industries and scenarios. vCloud NFV software components can be used in a variety of ways to construct a comprehensive, end-to-end solution that meets the business goals of communication service providers. vCloud NFV leverages VMware vSphere for compute virtualization, VMware NSX Data Center for network virtualization, and VMware vSAN for storage virtualization, as well as leveraging the VMware vRealize Suite set of products for analytics, monitoring and Day 2 operations.

vCloud NFV offers choice in terms of which Virtual Infrastructure Manager (VIM) it supports, namely vCloud Director and VMware Integrated Openstack.

For more information on vCloud NFV, go to vmware.com/go/nfv.

VMware SD-WAN components such as the VMware SD-WAN Orchestrator, the VMware SD-WAN Gateways, and the VMware virtual SD-WAN Edge can all be instantiated as Virtual Network Functions (VNFs) on top of a vCloud NFV-based NFVI, as shown in Figure 10.1 below. Third-party VNFs can be instantiated as well on the vCloud NFV NFVI. These third-party VNFs can include functions such as Next Generation Firewalls, virtual routers, WAN optimization appliances, and so on. VMware SD-WAN Edges would be instantiated at edge locations (Central Offices or Points-of-Presence/POP), whereas gateways and orchestrators would be instantiated in NFVIs in centralized data centers.

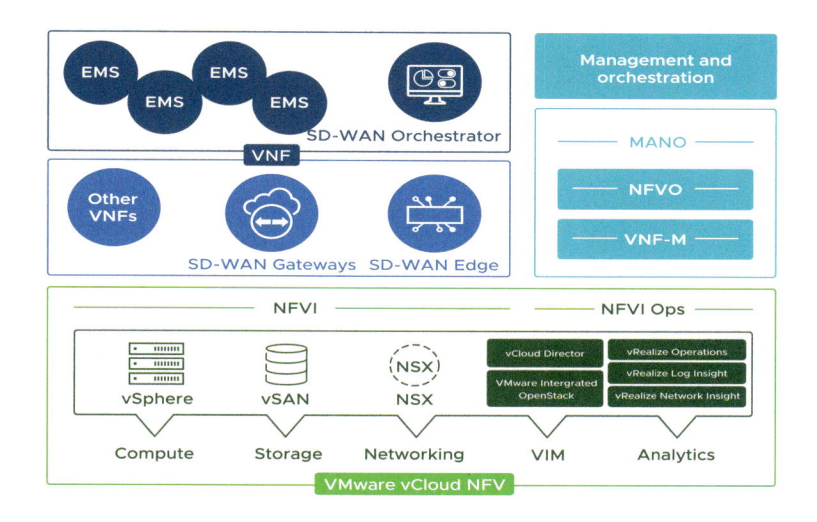

Figure 10.1: VMware SD-WAN VNFs on the VMware vCloud NFV infrastructure

Scenario: VMware SD-WAN Edge VNF in the branch

This SD-WAN deployment requires a VMware SD-WAN Edge instantiated inside a compute edge server at branch locations. To keep the cost of the compute edge low, as well as maintaining control over operational costs (such as electricity), low-powered servers are used in the enterprise branch. The compute edge needs to have enough resources for it to have the ability to add applications or functions to those already deployed as part of the service. The compute edge is often referred to as the universal Customer Premise Equipment or uCPE.

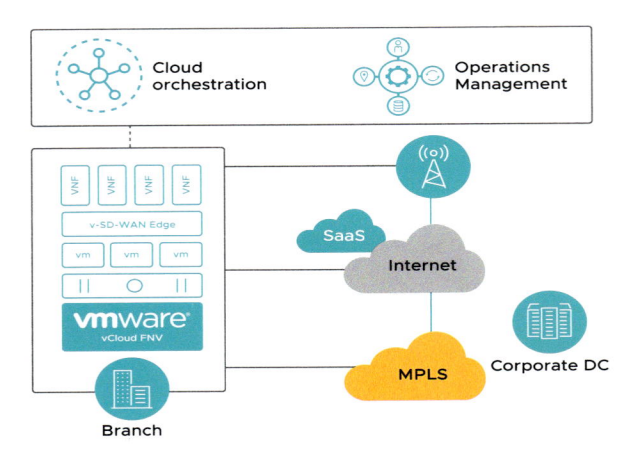

Figure 10.2: VMware SD-WAN as part of a universal CPE/Compute Virtualization in the branch

This design would be suitable for basic edge connectivity for branches with a moderate number of users. With local broadband loops and MPLS connectivity to the corporate DC, the solution provides the flexibility to support applications hosted locally, in the public cloud, and in the corporate DC, all in accordance with criticality, privacy, and security needs.

Branches will be managed from a regional or central headquarters' IT department, without the need for a network and security administrator at each branch. The branch edge will distribute connectivity based on business policies between direct internet access connections and the existing MPLS links to the service provider network. For example, corporate applications such as HR and Finance could be tunneled through the MPLS VPN circuits and/or 4G backup loops; cloud applications such as email, web collaboration, and sales management connected directly over the internet to VMware SD-WAN Gateways for Cloud on-ramp, and branch-to-branch video telephony could leverage branch-to-branch cloud-VPN using SD-WAN overlays.

Scenario: SD-WAN as mid mile/aggregation hub

In this scenario the VMware SD-WAN Edge(s) run inside the communication service provider's access/aggregation facilities such as Central Offices or Points-of-

Presence. This location serves as the connectivity hub for the branch offices and is likely to already have several IT components running VMware virtualization software. Furthermore this allows the solution to be overlaid on top of existing fiber to the home/business services. There is an VMware SD-WAN Edge deployed per branch, but this time in the Service Provider premise, not the customer premise.

Essential components to the healthy operations of an SD-WAN service, such as a VMware SD-WAN Gateway, and, depending on the service, value-added VNFs such as analytics engine, content caching, and IP-PBX are likely to be installed here. The aggregation hub approach allows enterprise customers to capitalize on the localized crowding and short-haul low-latency to connected sites.

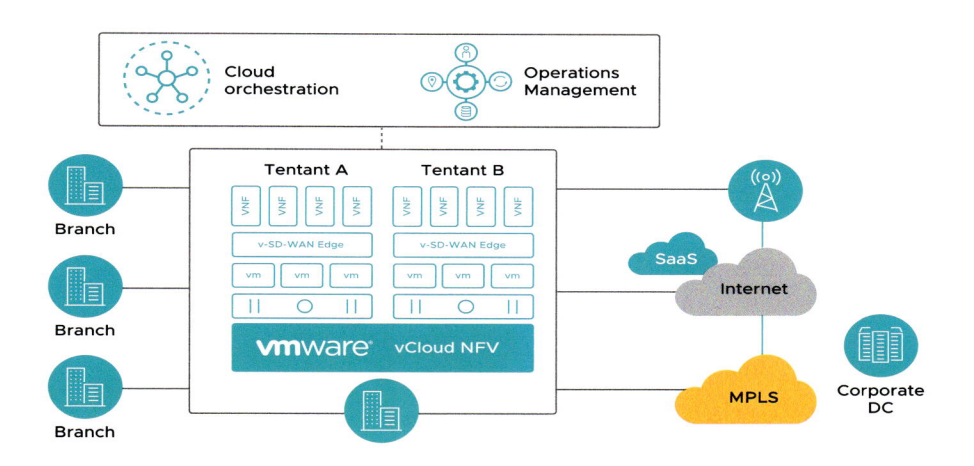

Figure 10.3: VMware SD-WAN as part of an aggregation layer

As the aggregation hub is owned by the communication service provider, vCloud NFV's multi-tenancy is essential, since several enterprises are likely to connect to this location, enabling branch-to-branch Cloud-VPN networking and mobile endpoints, among other services.

Scenario: Service Chaining of Value-Added Services

Communication service providers can quickly develop, provision and configure new value-added services at any site with seamless, centralized, cloud-based configuration and operations management. For example, communication service providers (or their enterprise customers) can rapidly enable such new services either at the branch, aggregation edge or corporate data center to meet their business and compliance needs.

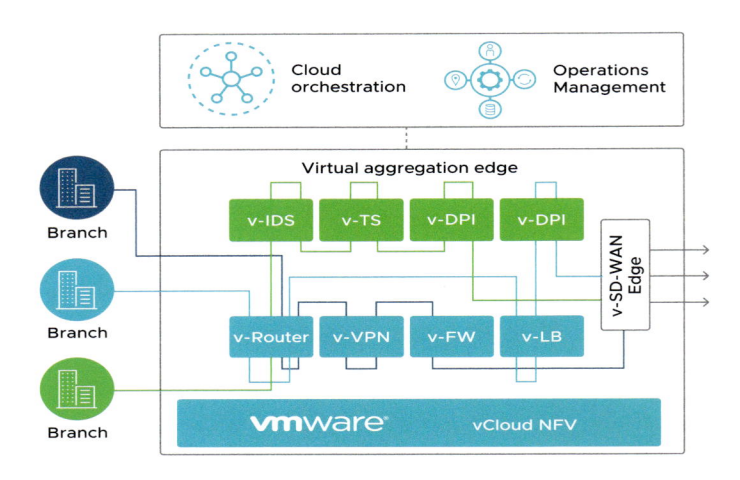

Figure 10.4: VMware SD-WAN and value-added services

The VNF onboarding and application composition is far simpler and automated. This allows a quick fast-fail trial or a quick production scale-out. The example illustration shows a virtual aggregation edge site with differentiated service compositions for each branch, ranging in content and traffic management, malware detection, intrusion detection, DPI and SIP trunks.

Future Use Cases

Some future use cases for NFV and SD-WAN could include:

- Edge Computing with NFV and SD-WAN

    - Mobile solutions for airplanes, trains, and cars.
    - Industrial remote IoT deployments such as oil well optimization, utility grids, and smart city use cases where the "thing"s reside in ruggedized, disparate, outdoor and often at times remote locations with inconsistent network and power.
    - Factory and plants in support of closed networks, ruggedized indoor environments; and branches, in stores in support of unique space and power requirements, and coordinated across many stores.
    - LTE/5G Network Slicing, Cloud Radio Access Network (RAN).

11. Appendices

Acronyms

2FA - Two Factor Authentication

4G - 4th Generation Cellular Network Technology

5G - 5th Generation Cellular Network Technology

ACL - Access Control List

AES - Advanced Encryption Standard

AI - Artificial Intelligence

API - Application Programming Interface

AWS - Amazon Web Services

BGP - Border Gateway Protocol

BSS - Business Support System

CA - Certificate Authority

CE - Customer Edge

CASB - Cloud Access Security Broker

CDE - Cardholder Data Environment

CLI - Command Line Interface

COS - Class of Service

CRLs - Certificate Revocation Lists

CSR - Certificate Signing Request

CSS - Cloud Security Service

DIA - Direct Internet Access

DMPO - Dynamic Multi-Path Optimization

DNS - Domain Name Services

DPI - Deep Packet Inspection

DSCP - Differentiated Services Code Point

DTLS - Datagram Transport Layer Security

ESP - Encapsulating Security Payload

FEC - Forward Error Correction

FQDN - Fully Qualified Domain Name

GRE - Generic Routing Encapsulation

GUI - Graphical User Interface

HTTP - Hyper Text Transfer Protocol

HTTPS - Hyper Text Transfer Protocol Secure

HW/SW - Hardware / Software

IaaS - Infrastructure-as-a-Service

IDS - Intrusion Detection System

IKE - Internet Key Exchange

IoT - Internet of Things

IPSec - Internet Protocol Security

KVM - Kernel-based Virtual Machine

LAN - Local Area Network

LTE - Long Term Evolution

MPLS - Multiprotocol Label Switching

MSP - Managed Service Provider

MTU - Maximum Transmission Unit

NAT - Network Address Translation

NETCONF - Network Configuration Protocol

NGFW - Next Generation Firewall

NFV - Network Function Virtualization

OSPF - Open Shortest Path First

OSS - Operational Support Systems

PAT - Port Address Translation

PCI - Payment Card Industry

QoS - Quality of Service

QoE - Quality of Experience

RADIUS - Remote Authentication Dial In User Service

RBAC - Role-based Access Control

REST - Representational State Transfer Technology

RSA - Rivest–Shamir–Adleman

SA - Security Association

SaaS - Software-as-a-Service

SDN - Software-defined Network

SD-WAN - Software-defined Wide Area Network

SHA - Secure Hash Algorithm

SLAs - Service Level Agreements

SNMP - Simple Network Management Protocol

SSO - Single Sign On

TCP - Transmission Control Protocol

TLS - Transport Layer Security

TTM - Time to Market

uCPE - Universal Customer Premise Equipment

UDP - User Datagram Protocol

URL - Uniform Resource Locator

VCMP - VeloCloud Control Multi Path Protocol

VCN - Virtual Cloud Network

VCO - VMware SD-WAN Orchestrator by VeloCloud

VCG - VMware SD-WAN Gateway by VeloCloud

VCE - VMware SD-WAN Edge by VeloCloud

VCC - VMware SD-WAN Controller

VIM - Virtualized Infrastructure Manager

VMs - Virtual Machines

VNF - Virtual Network Function

VPC - Virtual Private Cloud

VPN- Virtual Private Network

vRNI - vRealize Network Insight

VRF - Virtual Routing and Forwarding

VRRP - Virtual Router Redundancy Protocol

WAN - Wide Area Network

WPA-ENT - Wi-Fi Protected Access Enterprise

ZTP - Zero-Touch Provisioning

Glossary

3G/4G/LTE: 3rd / 4th generation and long-term evolution mobile / wireless communications network.

5G network: 5th generation mobile / wireless communications network denotes the next major phase of mobile telecommunications standards beyond the current 4G standards. 5G has speeds beyond what 4G can offer.

Access Control Lists (ACLs): simple network traffic filter to permit or deny packets based on packet elements such as IP address and TCP/UDP port.

Advanced Encryption Standard (AES): symmetric block cipher or encryption algorithm, implemented in software and hardware throughout the world to encrypt sensitive data.

Artificial Intelligence (AI): an area of computer science that emphasizes the creation of intelligent machines that work and react like humans.

BGP community: the BGP community attribute is a numerical value that can be assigned to a specific prefix and advertised to other neighbors. When the neighbor receives the prefix it will examine the community value and take proper action whether it is filtering or modifying other attributes.

Blackout network condition: a network that stops forwarding all traffic.

branch: an office location, other than the main office.

Brownout network condition: a network, which has a limited ability to forward traffic due to packet loss caused by a fault or congestion and typically seen as a valid forwarding path by traditional routing equipment.

Burst Mode: a VMware SD-WAN Edge link speed measurement option where link speed is measured at a high bitrate early in the test rather than slowly ramping up the bit rate.

Business Policy: a VMware SD-WAN rule set, used to apply link steering, QoS prioritization, NAT and service class settings to specific traffic types. Traffic type can be selected in the **Match** part of the business policy rule by matching on source or destination IP, port etc. and/or application signature.

Certificate Authority (CA): internal or third-party trusted entity that creates, signs, and revokes digital certificates that bind public keys to user identities.

cloud-hosted gateway: a VMware SD-WAN Gateway that is hosted on the public internet and provides multiple SD-WAN services such as control-plane routing, optimized data plane path to SaaS, IPsec tunneling to third-party security services and firewalls.

clustering mode: multiple VMware SD-WAN Edges configured as a single entity to provide a VCMP tunnel endpoint where new incoming tunnels from remote VMware SD-WAN Edges are load-balanced across Edges that are members of the cluster.

Command-Line Interface (CLI): text-based user interface (UI) used to configure and monitor networking systems.

Continuous Monitoring: a feature of VMware SD-WAN Dynamic Multi-Path Optimization (DMPO) which provides continuous network link and path monitoring in real-time.

control plane: logical description of a network function that provides signaling and routing.

controller: a VMware SD-WAN Gateway with data plane services disabled.

data center interconnect (DCI): technology used to link two or more data centers so that the facilities can share resources in L2/L3.

data plane: logical description of a network function that provides transportation of user and application traffic.

deep application recognition (DAR): a VMware SD-WAN Edge-based service that supports classification of traffic based on Layer 2 to Layer 7 attributes and recognizing over 3,000 applications through deep packet inspection and application signatures.

degradation: a condition where link or network performance and throughput is reduced i.e. degraded, due to packet loss. Packet loss could be as a result of a fault or congestion.

disaster recovery (DR): an area of security planning that aims to protect an organization from the effects of significant negative events. DR allows an organization to maintain or quickly resume mission-critical functions following a disaster.

dual-arm: a system with at least two connections between two networks.

Dynamic Application Steering: depending on the application or marking the traffic through, the VMware SD-WAN Edge will be forwarded to a special link, rate limited, or marked.

Dynamic Multi-Path Optimization (DMPO): a method employed by the VMware SD-WAN solution to deliver assured application performance and a uniform QoS mechanism across different transports.

Full qualified domain name (FQDN): is the hostname plus complete domain name.

gateway: see **cloud-hosted gateway** and **partner gateway**.

geolocation: estimation of the real-world geographic location of a device based on its public IP address.

goat yoga: a yoga instructor teaches class as they normally would, except baby goats have free rein to come and interact with you during your practice.

hash-based message authentication code (HMAC): is a specific type of message authentication code (MAC) involving a cryptographic hash function and a secret cryptographic key.

High Availability Mode: two VMware SD-WAN Edges configured to be highly available such that one Edge is active and the other is in a standby state, monitoring the active Edges via a heartbeat cable. The Edges in high availability mode share the same configuration and will show up in the VMware SD-WAN Orchestrator as a single device.

Hub: explicit role assigned to a VMware SD-WAN Edge which is typically located in the data center and which terminates multiple overlay tunnels.

Hub-and-Spoke VPN: a VMware SD-WAN hub-and-spoke topology describes one or more branch VMware SD-WAN Edges aka **spokes**, build one or more permanent tunnels to a centrally located edge in head office or the data center—known as the **Hub.**

internet backhaul: an option in a VMware SD-WAN Business Policy rule to redirect traffic that would normally be sent to the internet, via a hub edge or cloud-based security service (either via the VMware SD-WAN cloud-hosted Gateway or direct via an IPSec tunnel).

internet key exchange version 2 (IKEv2): standard for secure key exchange between peer VPN devices, as defined in RFC 5996.

Internet of Things (IoT): extension of internet connectivity into physical devices and everyday objects.

IPsec encryption: internet protocol security (IPsec) is a set of protocols that provides security for Internet Protocol. It can use cryptography to provide security. IPsec can be used for the setting up of virtual private networks (VPNs) in a secure manner.

Jitter: packet delay variation.

Layer 3: third layer of the Open Systems Interconnection (OSI) Model, known as the network layer and describes the layer responsible for all packet forwarding between intermediate routers.

on-premises deployment: hardware or software deployed at an organization's site rather than at a third-party location to the organization.

Over The Top: a VMware SD-WAN deployment into an organization leveraging the VMware cloud-hosted Orchestrator and Gateways rather than the organization hosting those components themselves.

overlay: an arrangement of virtual private network connections (VCMP tunnels in the case of VMware SD-WAN) built on top of an existing network (underlay), controlled by VMware SD-WAN Gateways and Orchestrator.

Overlay Flow Control: a feature developed for the VMware SD-WAN solution to provide a single pane of glass for visibility and control of the network-wide routing.

packet loss: refers to one or more packets traversing a data network which do not make it to final destination due to network congestion or other errors encountered.

partial mesh: refers to VPN configuration where not all nodes in a VPN are connected to each other.

partner gateway: VMware SD-WAN control plane and data plane component managed by a partner and typically deployed with both a public internet-facing and private network-facing interface for hand-off to different MPLS VRFs.

point-to-point services: refers to the private network connections directly between organization's sites.

port numbers: used by layer 4 protocols such as TCP and UDP to identify the client and server application session.

private circuits: permanent communication links dedicated for an organization's exclusive use.

private underlay network: refers to a connection such as MPLS or a leased line.

private WAN overlay: a VMware SD-WAN user-defined overlay that is carried over a private network where a VMware SD-WAN Gateway is not reachable.

public cloud: hardware and software services provided by third-party providers over the public internet.

Public Key Infrastructure (PKI): a technology to facilitate the secure electronic transfer of information for a range of network activities such as e-commerce, internet banking, and confidential email. PKI includes a set of roles, policies, and procedures needed to create, manage, distribute, use, store and revoke digital certificates, and manage public-key encryption.

Quality of Service (QoS): a networking framework or methodology used to protect critical application traffic during network congestion.

Rate limiting: in the context of Quality of Service, used to limit the bandwidth access for applications.

real-time: refers to applications such as voice and video with strict latency and loss requirements.

redundancy: it is the ability for alternate hardware or software to take over normal network operation in the event of failure of the primary.

Roles-Based Access Control (RBAC): the method of restricting network access based on the roles of individual users within an enterprise.

SaaS applications: Software-as-a-Service applications are usually web-delivered applications such as Office 365.

segmentation: a feature of VMware SD-WAN that allows individual LAN side interfaces or VLANs to be placed into one or more isolated routing tables on the VMware SD-WAN Edge and the isolation of traffic belonging to these networks is maintained over the WAN.

Slow Start: VMware SD-WAN-specific bandwidth measurement option, which allows measurement of the actual bandwidth of a connection. This method is the default for wired network connections detected in the VMware SD-WAN Edge GE ports.

Smart Defaults: the VMware SD-WAN default business policy configuration.

split brain protection: in redundancy configurations, where a standby device keeps the same state as the active device, situations can arise where the communication between active and standby is broken and both devices try to become primary for the communication going forward.

Spoke: in a Hub-and-Spoke network environment the spokes are usually the end devices or branches where end user devices are attached. They usually communicate between each other over the central Hub elements.

squirrel: members of the family Sciuridae, a family that includes small or medium-size rodents. The squirrel family includes tree squirrels, ground squirrels, chipmunks, marmots, flying squirrels, and prairie dogs amongst other rodents. Well known for their bushy tails.

Steve Woo: VP Products and Co-Founder of VeloCloud, now VMware.

The Big Lebowski: Crime/Indie movie by Ethan and Joel Coen from 1998 about a name mismatch and the consequences between an ordinary guy and a criminal. Involves a rug that "ties the room together."

Time-stamping: used in the VMware SD-WAN solution on every packet between data plane devices to track network delay in both directions and measure network latency in real-time.

trunk: usually refers to an uplink interface summarizing multiple data streams and individual virtual circuits on one physical link.

universal Customer Premises Equipment (uCPE): a device with switched and routed ports used to combine a multitude of networking services on a single hardware platform.

underlay: existing network infrastructure used by VMware SD-WAN to carry overlay traffic (VCMP tunnels) and can be public (internet) or private (MPLS, point-to-point etc.).

vCloud NFV: vCloud NFV is a modular, multi-tenant Network Functions Virtualization platform featuring compute, storage, networking, management and operations capabilities from VMware.

virtual cloud network (VCN): A ubiquitous software layer from data center to cloud to edge infrastructure from VMware.

Virtual Routing and Forwarding (VRFs): concept in router elements in a network to host customer specific routing information separately from each other on the same device. They also allow customers to use overlapping address spaces.

VLAN ID: located on layer 2 of the OSI model for networks, the VLAN ID is carried in an 802.1q frame and is used to define virtual channels on a layer 2 (usually Ethernet) network segment.

vSphere: hypervisor software from VMware to virtualize x86 server hardware in data centers.

zero-touch provisioning: marketing term describing the configuration of devices without manual intervention at the location of deployment, apart from unboxing, racking it, cabling it, and triggering activation.

Additional Resources

For additional VMware SD-WAN by VeloCloud information or resources, visit our dedicated resource page (*http://split.to/J4hmnsv*) and the documents below.

- VMware SD-WAN Homepage: *https://www.velocloud.com*
- VMware SD-WAN Datasheet: *http://split.to/Q15Fg83*
- VMware SD-WAN Communities: *http://split.to/iUB2Z6r*
- VMware SD-WAN Blog: *http://split.to/JKNP7sV*
- VMware SD-WAN Support Portal: *http://split.to/oiZkeUr*
- VMware SD-WAN Orchestrator API: *http://split.to/ME0jHPM*
- VMware SD-WAN Hands-on Lab: *http://split.to/UdrWiLO*
- VMware Knowledgebase: *http://split.to/xn3bLZo*

For additional and future VMware SD-WAN information or resources, please visit the following:

- VMware vCloud NFV: *http://split.to/hVi1iL8*
- VMware VNC: *http://split.to/3pSnDEu*
- VMware Cloud Foundation: *http://split.to/qM3oIzw*
- VMware Edge and Internet-of-Things (IoT):*http://split.to/GudFjRb*

For general VMware information or resources, please visit the following: VMware: *https://www.vmware.com/*

Colophon

Fonts: Open Sans by Steve Matteson and Metropolis by Steve Matteson

Book Sprints (*www.booksprints.net*) team:

- Barbara Rühling (Facilitator)
- Henrik van Leeuwen (Illustrator)
- Agathe Baëz (Book producer)
- Raewyn Whyte and Christine Davis (Copy editors)

Warning & Disclaimer

Every effort has been made to make this book as complete and as accurate as possible, but no warranty or guarantee is implied. The information provided is on an "as-is" basis. The authors, VMware, and the publisher shall have neither liability nor responsibility to any person or entity with respect to any loss or damages arising from the information contained in this book.

The opinions expressed in this book belong to the authors and are not necessarily those of VMware.

VMware, Inc. 3401 Hillview Avenue Palo Alto CA 94304 USA Tel 877-486-9273 Fax 650-427-5001 *www.vmware.com*

Copyright © 2019 VMware, Inc. All rights reserved. This product is protected by U.S. and international copyright and intellectual property laws. VMware products are covered by one or more patents listed at *http://www.vmware.com/go/patents.* VMware is a registered trademark or trademark of VMware, Inc. and its subsidiaries in the United States and/or other jurisdictions. All other marks and names mentioned herein may be trademarks of their respective companies.